Praise for *Behind the Binoculars*

"this engaging volume provides a unique insight into what makes birders 'tick'. …a valuable reflection on the development of birdwatching over the best part of the last century."
—NIGEL HOPPER, *Ibis*

"There are many fascinating tales in this collection of birders' biographies, some of which have passed into birding folklore."
—ADRIAN PITCHES, *British Birds*

If you are looking for a treatise on the private life of the birder then take a peek at the supremely entertaining and highly informative *Behind the Binoculars*.
—STUART WINTER, *Sunday Express*

"thoroughly entertaining and informative interviews."
—MATTHEW MERRITT, *Birdwatching Magazine*

"For me this was a nostalgic book. I related closely to many of the stories, particularly by the older people. It was a fun book to read and should appeal to all avid birdwatchers. So take it to the cottage and curl up by the fire when the rain is heavy and enjoy."
—ROY JOHN, *Canadian Field-Naturalist*

"This passionate book does more for conservation in one reading than many dry academic papers on the dangers that surround our birds today."
—CERI LEVY, *Caught by the River*

"This is both a serious overview of the field and a flock of delights, from the shot of a youthful Betton with three young song thrushes balancing on his forearm to fond memories of first binoculars, whether Leica Ultravids or Swarovskis."
—BARBARA KISER, *Nature*

"I was fascinated by what initially sparked each person's interest in birds and lit a fire that lasted a lifetime – it seemed to me that in all cases they were young and, like us all at that age, had very dry tinder!"

—**DEBBIE TODD,** *www.bto.org/about-birds/book-reviews*

"The real common thread… is a love of birds and a determination to make a life devoted in some way to their enjoyment, study or conservation. All the interviewees have found the same early fascination but have then moulded it according to their talents, circumstances and opportunities to make a genuine and lasting contribution. Birds, it would seem, can take you anywhere."

—**ANDY STODDART,** *Rare Bird Alert*

Behind the Binoculars

Behind the Binoculars
Interviews with acclaimed birdwatchers

Mark Avery and Keith Betton

PELAGIC PUBLISHING

Published by Pelagic Publishing
www.pelagicpublishing.com
PO Box 725, Exeter EX1 9QU, UK

Behind the Binoculars: Interviews with acclaimed birdwatchers

ISBN 978-1-78427-145-9 (Pbk)
ISBN 978-1-78427-051-3 (ePub)
ISBN 978-1-78427-052-0 (Mobi)
ISBN 978-1-78427-053-7 (PDF)

British Library Cataloguing in Publication Data
A catalogue record for this book is available from the British Library.

Cover image: *Turnstones* by Robert Gillmor

Printed in India by Imprint Press

This book is dedicated to the memory of three great birders, Roger Tory Peterson, Guy Mountfort and Phil Hollom, whose field guide opened the eyes of many of us to the world of birds.

Phil Hollom was interviewed for this book but died, aged 102, before its publication.

Contents

Preface

We are both passionately interested in birds – and quite interested in people too. Bringing these two interests together in a book where people interested in birds talk about how they got interested, and what they think of birds, bird conservation and other birding people seemed quite a fun thing to do.

Over a bottle of wine and a pizza in a restaurant in Victoria Street, London, we decided to carry out this project together and after a search for a publisher and numerous lists of potential interviewees we finally settled on the list of eminent birders whose stories and views are chronicled here.

Choosing our interviewees

We aimed to interview a selection of people who would recognise themselves as being birders – people for whom seeing birds has been a major part of their lives. For all, birding has been the most important and engrossing pastime and for some it has also been central to the way they have earned a living.

We were looking for a variety of standpoints and opinions.

Our interviewees range in age from the centenarian Phil Hollom, to a child of the 1970s, Rebecca Nason. Two were born in the 1930s, two more in the 1940s, five in the 1950s and a further nine in the 1960s.

Sixteen of the twenty interviewees are men.

The interviews

One or other of us talked to each of the people in this book, and the interviews were transcribed, edited and then approved by the people we interviewed.

The earliest interview was with Phil Hollom, carried out in 2009, but most were carried out in 2012–14. Several interviews have been updated as time has passed, but we had to call a halt at some stage and so events occurring after mid-2013 are generally not covered here.

Spoken English is very different from written English, and we have tried to retain the conversational nature of the interviews but also make them relatively easy to read.

There are a few terms scattered through the interviews which might perplex some readers – most are to do with the field sport of 'twitching'. Not all, in fact rather few, birders are twitchers even though the popular media can't seem to get that fact straight. 'Twitchers' are those who rush after rare birds hoping very much to see them. They twitch with a mixture of excitement and nervousness when they hear of a rare bird that they wish to see. If they fail to see it, then they have 'dipped', and if others see a rare bird which they miss, they have been 'gripped off'. If twitchers see a bird they've never seen before it is, at that time, a 'lifer' as it has been added to their 'life list' of birds (and by definition their 'year list' too). Anyone who develops a reputation for unreliable or unverified sightings of rare birds may, behind their back, be labelled as a 'stringer'.

Many British birders have 'British lists', some have 'county lists' and others might have 'garden lists' – some have longer lists of lists. 'Blockers' are species that are very rarely seen, but that others saw and you didn't, so that your ability to surpass their total number of birds seen is hampered by those, perhaps irretrievable species. Just as Marvin Gaye 'heard it on the grapevine' that is the phrase used by birders, for their network of contacts who pass on information on interesting birds locally or nationally – in the days before the internet, pagers and mobile phones it was even more important to have a good network of contacts. A bird's jizz is its general appearance, shape and behaviour – rather similar to 'gestalt'. Many birders do much of their regular birdwatching at a regular spot, often close to home, known as their 'patch'. If a lot of migrating birds arrive overnight, usually on the coast and because of poor weather, then it is said to be a 'fall' of migrants (whether it is spring or autumn).

The order in which the interviews appear in this book bears no relation to the order in which they were done – or to anything else, except to what we thought was a good order.

When we paused for breath, with all the interviews completed, we saw that we had been lucky (maybe skilful?) in picking some fascinating people with a wide range of backgrounds and perspectives. We also realised that this book takes the reader behind the binoculars of famous birders and into their heads, their thoughts and emotions.

Each interview stands alone as an interesting account, but there are also some common themes, or differences, that leaped off the pages. We discuss these in the last chapter.

Mark Avery and Keith Betton

May 2015

Acknowledgements

We are grateful to Nigel Massen at Pelagic Publishing for encouraging us to put this book together; to our interviewees for their openness and patience; and to our partners (Esther Betton and Rosemary Cockerill) for their patience and help.

The interviews with Phil Hollom and Ian Wallace are updated versions of ones which first appeared in *Birdwatching* magazine – we are grateful to *Birdwatching* for permission to reproduce them here.

Keith would also like to thank his sister, Alison, for lending him her *Observers' Book of Birds* in 1968 – otherwise he would probably have become a train spotter instead of a birder!

CHRIS PACKHAM

Chris Packham was born in the 1960s and is best known as a nature photographer, TV presenter and author.

INTERVIEWED BY KEITH BETTON

What was your first experience of birds?

For my second birthday my grandmother bought me two books of British birds, with rather colourful pastel illustrations – I still have them, they're dedicated to me: *From Granny and Granddad, 1963*. I can still remember every single illustration – my favourite was the Merlin, and I also liked the Hobby and Wheatear. Looking back at them, the illustrations were very friendly – the birds had quite large eyes and were attractive – but I think what drew me to them was their diversity. One was *Birds of Heath and Marshland* and the other was *Farmland Birds*. I was into masses of other animals; I was into things you could touch and I had to put them in a box or a jam jar and of course I couldn't do that with birds. It was tadpoles and ladybirds, then slow-worms and lizards, then snakes – I had a huge reptile collection by the time I was eight. But I remember being at a bus stop and there was a dead Starling in the gutter. I picked it up, and my mum said, 'Don't touch that dirty thing'. Obviously not listening to what my mum said, I held it in my hand and unfolded its wings and I think that's when I fell in love with birds – because it was perfect. The ladybirds had been perfect, the slow-worms had been perfect, but the patterning and overlaying feathers were just exquisitely beautiful. I remember stroking the feathers on its head and looking at the iridescence on its nape and I was just entranced by it. I still have its wings at home.

After that, like many other people, I started collecting birds' eggs at the age of eleven. I found my first nest, a Chaffinch's, in the grounds of Bitterne Park Secondary School, and I took and blew one of the beautiful marbled red-and-cream eggs. I had *The Observer's Book of Birds' Eggs* and I collected for a couple of seasons. At the same time my dad had bought me the Heinzel, Fitter and Parslow guide (*The Birds of Britain and Europe, with North Africa and the Middle East*) and my first pair of binoculars, which were from Boots (really shocking ones – they didn't last long, as they used to bounce against the handlebars of my bike when I had them round my neck and so they soon got smashed to pieces).

The first bird I remember identifying from the Heinzel, Fitter and Parslow guide was a male Baltimore Oriole at Gaters Mill on the River Itchen. I was on my bike, looked up and saw it, cycled home furiously, burst into the lounge where my dad was watching *Grandstand* and said, 'Dad, look at this, I've spotted one of these!' My dad, in his usual calm, collected way, took the book from me, looked at the bird, passed the pages through his fingers till he got to Grey Wagtail, and pointed out that it was far more likely to be that than a Baltimore Oriole in summer plumage.

I became an obsessive nest-finder. I only collected one egg from each species and started to map them in the diaries which I'd been keeping – I mapped all the nests, counted all the eggs. Then, one lunchtime, when I was twelve, I was speaking to John Buckley who was my biology teacher and he asked what I was doing. I told him about the nests and showed him my diary with all the maps. He pointed out that collecting the eggs was a waste of time and that the data was far more important. So the egg-collecting stopped immediately – I still went nesting a lot and wrote down all the information, and he would come and ring the young, so it was all about counting the nests, counting the young – and I couldn't wait for one of them to die so that we had a recovery! I remember we once ringed eight Little Owls in a nest on Itchen flood plain and one of them got handed in, having been shot by a keeper, who thought it was a rabbit coming out of a burrow. I also went out mist-netting birds with John Buckley, and holding my first Treecreeper in my hand in Decoy Covert on the Lower Itchen was amazing – it was at that stage where touching a bird was important to me.

Then I got massively into Kestrels, and I gave up mapping other birds' nests and just mapped Kestrels. I had a study area of 27 square miles, stretching from St Mary's College in Southampton to Winchester Cathedral and I was determined to find every Kestrel nest in this area. I did that for three or four years.

Then I took a Kestrel from a nest on 25 June 1975, I trained it and flew it. There were all sorts of shenanigans about this – I applied for a licence from the Home Office (as it was then), because at that time they were granted to people to take birds from the wild – but not to working-class oiks from Southampton. Falconry was still the preserve of gentlemen, even when it came to Kestrels. I'd sought the help of other falconers, notably Phillip Glasier. I asked for his help but he was incredibly nasty and the way he treated me at the age of fourteen had a profound effect on me, in that the way I behave towards fourteen-year-olds is the polar opposite – I will do anything to encourage their interest, not everything to destroy it. In the end, though, he actually made me more determined.

When I had the Kestrel I had all sorts of problems; for instance, the Royal Society for the Prevention of Cruelty to Animals (RSPCA) wanted to take it away. Owning that bird was a formative period in my life because it really turned me against – and sowed the seeds of a total disrespect for – authority. It isolated me from almost everything: I was adolescent, but was massively into wildlife when other kids were going to parties and hanging round the ice rink. I just became an outsider and hung around on my own. The bird died on 6 December and that also had a profound effect on me, because I didn't get a lot of emotional support getting over it and I was very, very distressed about losing it. My father was – perversely – in favour of me having the bird: he didn't like the fact that I'd taken it illegally, but he recognised that it was important to me. He was a very matter-of-fact, letter-of-the-law kind of guy, but at the same time he recognised that it was unfair that I wasn't able to do it legally and knew that I was going to do it anyway. When it died, I was so distressed that I couldn't speak for a week, just couldn't make words come out of my mouth – my parents sent me to school but it was a disaster, as I couldn't articulate my distress and had no way of getting over the trauma.

So it's fair to say that, apart from your mum and dad, that's the closest relationship you had at that point in your life?

Yes. I've had other animals since then and had very intense relationships with them. They don't lie to me – my dogs now are the epicentre of my universe.

So the bird died, and then punk rock came along, which was brilliant as part of it was about alienating yourself from other people. In terms of birding, the Kestrel study continued; I won a Prince Philip zoology prize in 1979 – John helped me write it up; and I gave a talk at the Edward Grey Institute Conference in 1978, which was one of the highlights of my life. I'd sent them some of the findings of my Kestrel study and they invited me along – they didn't know who I was, of course, so when I turned up in trousers full of zips, a leather jacket and torn T-shirt and sat in the front row I could see the horror written all over their faces. Then the most brilliant thing happened. My birding hero at the time was Ian Newton, who'd written *Population Ecology of Raptors*. After I'd finished my talk everyone politely clapped, and Ian leaped up as I got off the stage, came over and shook my hand, and said my talk had been brilliant, and that meant so much to me, as I'd been totally alienated by just about everyone, but he could see past whatever colour my hair was, my trousers and studded biker jacket!

Who was your biggest early influence?

Definitely John Buckley.

What was your first telescope?

A Bushnell Discoverer 20–60 zoom, bought for me in 1979 by Joanne, my first girlfriend. I was at college. She had a job, saved up for it, bought it for £163.50 at the London Camera Exchange and gave it to me for Christmas. I still own it and wouldn't part with it. I don't have many sentimental things, but I was eternally grateful to her for that scope, which radically changed my life as I could now see birds properly!

Did you join the Young Ornithologists' Club (YOC)? [The precursor to the RSPB's Wildlife Explorers.]

No; I was really anti-social as a kid. My parents tried to get me to do lots of things to expand my interests socially, so they took me to the Southampton Natural History Society (of which I'm proud to be President). There I was very pleased to meet and mix with people in their fifties to eighties, as they were a fount of knowledge and I remember really enjoying going on some of their forays. There was David Goodall, who worked with bats and lived in Otterbourne and became a lifelong friend, and Dr J.G. Manners who was a fungal expert (who I joined up with again at university). So I was happy mixing with older people but didn't want to mix with kids my own age.

Did you join the Royal Society for the Protection of Birds (RSPB)?

No, I didn't join until probably the early 1980s – I remember having a membership card when I was at university, and visited reserves such as Titchwell and Minsmere. I was a bit mistrustful of all those sorts of people, and for me it was all very much an individual thing, me going out on my bike with my binoculars. It wasn't something I ever shared.

And then you finished school and went to university?

I went to Richard Taunton Sixth Form College in Southampton, where I studied biology under Alec Falconer, who was crucially supportive. I was a full-on punk at this time and some of the staff were trying to get me expelled for looking like a freak and refusing to play football. I played football for the school but refused to play for the college because I was taking five A-Levels and didn't have time. Alec kind of looked after me there, otherwise I think they'd have found an excuse to throw me out. I was frequently getting beaten up for looking like a freak, and it was a tough time. I enjoyed my secondary education greatly, and university tremendously, but the bit in between was very tough.

Where did you go to university?

Southampton. I'd given up Kestrels and started working on Badgers. Colin Tubbs, who I'd met years before thanks to Buzzards, offered me a project to look at the population of Badgers in the New Forest because I'd already been working in the Itchen Valley. I was two years into that and getting some really good data, so I went to Rory Putnam at Southampton University because there was quite an active mammal ecology group there. I showed him the work I was doing and said I'd like to continue, as the Nature Conservancy Council were potential employers – that's who I wanted to work for and I didn't want to let Colin down and dump all this data. I wanted to finish the study and so the university made me an unconditional offer – they must have wanted me to go there, so that was fabulous.

When I left home, I'd go back every Sunday morning and meet my father at an agreed time, just before it got light, and we'd go out birding. He wouldn't say very much, but he and I would go out, walk up the Itchen, drive to the New Forest, find a few nests, see a few Otters, etc. My dad liked walking, being outside, history, context and place, and we'd just amble about and if he saw something he'd point it out. I don't remember us having many conversations, because we had little in common at that time other than going out birding.

When I got to university my sister's French teacher, Dave Scott, who was at the school I'd attended (he didn't teach when I was there), became very keen on birds. He was an extraordinary character and the antics we got up to were phenomenal; worthy of a book on their own. He was a very colourful eccentric, but obsessed. Every single Saturday morning, without fail, no matter what state he or I were in hangover- or sleep-wise, he'd rock up at Joanne's house, where I was living, and we'd go birding. We went out in his tatty crumbling Triumph Vitesse, and our birding range was between Portland and Selsey Bill.

Was this twitching or birding?

We birded, but obviously if there was something special about we'd go and look at it. I remember at migration time seeing a Savannah Sparrow at Portland (obviously ship-assisted), but mainly we were after Honey Buzzards, Hawfinches and Red-backed Shrikes (which were still nesting in the New

Forest at that time). I could find nests better than most people at that point, but that's when I learned my skills as a birder – song, shape, everything else. Outside of a few key species – Kestrels, Sparrowhawks, things that I'd been obsessive about – in terms of general birding at that point I probably hadn't looked twice at Garden Warblers, Lesser Whitethroats or Whitethroats and identified them. I used to obsessively read *British Birds* and *Bird Study*, the British Trust for Ornithology (BTO)'s magazine, between the ages of eighteen and twenty-two.

But you didn't join the Hampshire Ornithological Society?

I wanted to do something with Great Grey Shrikes. I'd been working with Red-backed Shrikes in the New Forest, mapping them and on a personal crusade to try to stop egg collectors – I used to spend all day at the nest. I lost that battle, but then I got a grant to do some radio tracking on Great Grey Shrikes. I approached John Clark and asked if I could see the Great Grey Shrike records, which were kept in the Southampton University Library, which he happily agreed to. I then got a letter from Eddie Wiseman saying I couldn't look at them. I spoke to Colin Tubbs and asked him to have a word with Eddie, which he agreed to do, but he reported back to me that Eddie was adamant that he didn't want me to have access to the Hampshire Ornithological Society records. This just fuelled my dislike for a collective of people who I saw as a clique – I couldn't understand why Colin was so generous. John Clark sent me a very polite letter back, but this bloke I didn't even know was being so unhelpful. So I just thought I'd just do my own thing.

And then you got your degree?

My degree was in zoology. In 1982 I had a PhD offer from Oxford University to study Marsh Warblers. My tutor said I should apply for it as I'd definitely get through to the interview stage and it would be good for me. I went up to Oxford on the train and I sat in the library waiting, reading an article about sharks, when a woman came up and put her hand on my shoulder and said, 'Excuse me, but I'm going to have to ask you to wait outside; you're disturbing the other people in the library.' This was simply because they didn't like the way I looked. I went and sat outside. I was called into an office where an academic scowled at me and said, 'Do you have a pen?' I said yes, and he said, 'Do you have a piece of paper?'

I said no. He handed me a piece of paper and said, 'Write down this formula,' so I wrote down the formula; a mathematical model I was vaguely familiar with. He said, 'I'm going to give you three sets of figures and I want you to do a calculation,' which I did, and I remember thinking that this algebra test was rather unusual and not the sort of interview I'd imagined. After a couple of minutes, he asked what my answer was. I told him I hadn't completed it yet and he said, 'I don't think you're the man for us – off you go.'

He struck me as a seriously, seriously unpleasant man. The difference between him and Ian Newton is massive. You don't treat young, impressionable, avidly driven, obsessive, ambitious zoology students like that, so I thought that, for all the endeavour I'd put into achieving something academically, maybe I don't want to be in a world which is going to take one look at my spiky hair and then abuse me like that. So I cancelled my PhD and gave it up.

Why don't you have blue hair now?

Takes too much time; too much faff. I thought about bleaching it back to something outrageous, like Andy Warhol, when I was fifty-odd. Maybe I'll do it when I'm sixty. I've never wanted to look like anyone else. I'm very comfortable with my appearance, and I don't feel I have to fulfil a stereotype of any kind. It's not confidence; it's that I don't care.

So you kept notebooks. Did you take photos as well?

I only started taking photos after I left university. I didn't have a camera of any merit. I'd photographed Kestrel nest trees on a Zenith E camera, but I didn't take any real photos until I got my first Canon A1 with a 500mm mirror lens, and I bought that in my last year at university by selling my punk band gear. Joanne and I went off to the Camargue for six weeks. We slept in the car and I took lots of photos – all rubbish! But I was so entranced by being able to make things out of light and I'd learned enough in making those initial mistakes that I really wanted to take photographs. So, in order to finance that, I got a job as a camera assistant with Stephen Bolwell. He was making films for the BBC and that was how I ended up doing some broadcasting. He and I were soul mates: hugely argumentative, hugely passionate about art, cars, literature and that sort of stuff, and we just bounced off one another. Although we met through having a ferocious argument, we never had another, and we still correspond. He was another great mentor in my life, tremendously supportive in terms of giving me work, and helping me financially and with equipment.

So you got the job as his assistant – how did your career develop then?

I got to know people at the BBC, made a few wildlife films for them myself as a cameraman, then in winter 1985 *The Really Wild Show* started – I tried quite hard to get an audition and even when I'd done the audition I rang them and they said they'd ring me back and never did, so I spent the last of my dole money on buying a train ticket to Bristol. Because of my film camera work I knew someone who could get me into the building, and I made an excuse to see the lady I'd been working as a cameraman for. After making small talk I went to see Mike Benyon, producer of *The Really Wild Show*, and said, 'Listen, I've got a life; I've got things to do. If you're not going to give me the job, just summon up the balls to pick up the phone and tell me to clear off – otherwise you're just wasting my time and I don't like it.' Luckily for me, he was exactly the right bloke to say that to, because he and I had similar confrontational attitudes, and he thought that was bloody marvellous and gave me the job.

I had to write and research my own pieces for *The Really Wild Show*. Someone rang me up and said, 'What do you need for this piece?' and I said, 'A tank and some seahorses' and they said, 'OK then,' and that was it. I had no training; I just turned up in Birmingham and there it all was, a camera and some seahorses, and I just had to get on with it. I asked how long the piece was supposed to last and Mike said, 'I don't know, three or four minutes? I'll cut it out if it's rubbish.' So I stood there and talked about seahorses for three or four minutes and they put it on TV – it was absurd but wonderful!

After a few years I gave up *The Really Wild Show* and did a few other programmes for the BBC, and a series about wildlife photography for Channel 4 and a couple of programmes for ITV. Then I went back to the BBC to do a series called *X Creatures* which was about crypto-zoology – I put a lot of time and effort into it, and wrote and presented it. Then I met Jo, and she had a daughter, Megan, who's now nineteen, and I wanted to see her growing up, so didn't want to spend most of my time out in the country. I set up an independent production company called Head Over Heels. It immediately became successful and we had more work than we could handle, but the work was all overseas. I did two series with Michaela Strachan in South Africa, which were shown on the Discovery Channel, and they wanted another series, but that was another six months out of the country – Megan was about three at that stage and I just said, 'No thanks – it's over.' Having children wasn't something I'd planned, but all of a sudden one had turned up in my life and I needed to be there for her.

So I got a job working for the BBC in Southampton doing wildlife stuff on a programme called *Inside Out*, and I spent a lot more time at home with Megan. We did a lot of travelling together – she's been all over the world: Antarctica twice, India, Africa and Europe – and I don't regret a minute of it. It's been a massively important part of my life. I don't regret giving up professional travelling – I missed it, but I travelled with her. Once I was standing with a cameraman called Alex Hanson in Kathmandu, looking over the sunset from Swayambhu. I said, 'It's beautiful, isn't it?' and he replied, 'Yes. It's such a pity we're not here with anyone we care about.' It's just not the same, so when I travelled with Megan, all of a sudden I was sharing the travelling with someone, not just collecting postcards that I couldn't even talk about to people at home.

I did another couple of series called *Nature's Calendar* and *Nature's Top 40*, which were low-budget daytime series, but again I was lucky to work with a really superb team. They were as good as any BBC Natural History Unit programmes, so maybe it made the BBC prick up their ears a bit – and that's when I was offered *Springwatch*.

Did you find presenting easy?

Very easy; I've never been nervous. But, equally, I don't want to waste the time of the cameraman and soundman and if you're working on live TV it's very important that you do 'take one' as it's going to be the only one. I'm ruthlessly self-critical so it's a matter of being professional, and I think the most important part of my job is being a pleasure to work with, so I carry the tripod, I make the tea, I go to the shop and get the muffins, and I turn up on time, presentable, sober and informed, and I take a lot of pride in that. I never want to let my team down.

But can self-critical people be difficult to work with?

Hopefully not, in my case! The best people I've worked with are ruthlessly self-critical, because we have a limited amount of time, a limited amount of money to do the best job we can, and we're in a team, so there are always limitations – so if it's not the way we work together, it's the environment, the time, the money we've got and everything else. Most of the time I'm lucky to work with people who want to give 120%, and collectively it's about the result, so it's not always about doing things your way; it's about offering what you can into the mix, but respecting other people's abilities. There's more than one way to skin a cat; things don't always have to be done my way and the results can still be good, and co-operating for a synergistic product is important.

Sometimes when I have watched *Springwatch* or *Autumnwatch* in the past it has seemed like a contest between the presenters about who was going to get in and say the most, and cutting across each other, etc. I hate that. It's not a problem for me – if someone wants to start talking, I'll just stop. I don't care if it's on TV or in the pub; I just won't talk over anyone. Kate [Humble] and I immediately had an understanding: we could look at one another and know when the other one wanted to speak, and we would stop. That's a professional skill, and you need it because otherwise it looks untidy and scrappy and rude – if one of my colleagues wants to say something more than I do, then I'm happy for them to do it. I have no ego in terms of my TV work – if I have an ego at all, it possibly comes in terms of my photographs, because I work incredibly hard to do different things with my photography. My job in TV is to communicate my enthusiasm, infused with information, to inspire an audience in an entertaining way. I need them to listen because I want to take them somewhere, and at the end I hope they have learned something; that their affinity towards wildlife is enhanced, and vocationally I hope that means that we've got more people on our side. And for me, *Autumnwatch* and *Springwatch* are vocational programmes – when they offered it to me I was about to launch into a really intense period of wildlife photography. The whole thing was set up, and I asked myself if I really wanted to change my plans. Then I looked at the interactive nature of the programmes, the huge amount of public activity on the internet, Flickr, all that sort of stuff, and I thought I might be able to add something, to engage with the audience in a slightly different way, and perhaps even expand the audience. Having seen the programme, one thing I thought I could do was bring more science into the mix, up the ante – I don't think things had been dumbed down, but I wanted to raise the bar, and I still fight hard to do that.

Is the team you have at the moment a good mix?

Of course, it's not all about the presenters: it's about a huge and talented team behind the scenes. I'm a sort of constant consultant, as I'm being phoned from *Springwatch* and *Autumnwatch* all the time. That is not a problem. I like the programme, I want it to be as good as possible, and I think we have a very strong core team at the moment. They too want more wildlife, more innovation and better programming, and I like working in that atmosphere.

Where do you think wildlife broadcasting will be in ten years' time?

I think that's quite an easy prophecy. I did a series a couple of years ago called *Secrets of our Living Planet*, which was an authored piece by me and involved global travel. About 70% of the funding for the series came from the National Geographic Channel and obviously that put some constraints on what we could and couldn't do. As it was, we didn't compromise on any integrity for the version that went out on BBC2, but nevertheless it had to be cut without me and sold as a blue-chip series to National Geographic – that's what they got for their 70% of the money.

The BBC licence fee can't just cover highly expensive wildlife and drama programmes – the drama programmes have to be sold on as DVDs across the world, and wildlife programmes the same, but principally from an early stage they are co-productions. The projection is that co-production money will have to increase, maybe up to 90%, so only 10% of those programmes will be funded by the licence fee. The fee has been frozen for five years and the BBC – which is one of the greatest institutions and companies that we have, globally renowned, with programming of the highest calibre – has the living daylights beaten out of it by all and sundry. And yet it is a bastion of public service broadcasting. Now we've got BBC4, which is a lot like the old BBC2: the budgets aren't high, but the programming is really good. I would hope that people think the BBC is something we need to value and fund. The licence fee is a piffling tax when compared to what people spend on newspapers, magazines or hiring DVDs, and to quibble over that is insane.

What do you think we should do now to stop this happening in ten years' time?

We polarise, because that's what's happening generally. There will be low-budget programmes and mega-budget programmes and less in between. The mega-budget mega-series will continue because they are principally funded from overseas – and that's not just one network in America. I think something like *Springwatch* will be more difficult, budget-wise, because it's a live programme that only goes out in the UK. It's a flagship brand for BBC2 and thankfully they don't want to lose it – and we're grateful that they see it that way – but it's an expensive show to make, and is all over very quickly. And they're not making money by selling DVDs or anything else. There's no merchandising. It's not branded to that extent, which is a pity.

Do you think programmes will be more celebrity-based, or will the wildlife still be the star?

If the wildlife isn't the star, we're lost. I personally don't like celebrity wildlife programmes, and nor do I like celebrity history, chemistry or art programmes. I want to listen to people who I can immediately sense are speaking from the heart and are doing it because they believe in it – I don't want to watch someone who has no interest in the subject, reading from an autocue, not for natural history or any of the specialist disciplines. I think the BBC realises that part of its integrity comes through employing people who are specialists – and I'm not saying that because I want a job and I'm a specialist! If we go back to the great broadcasters – David Attenborough, Alan Whicker, those sorts of people – you could tell immediately that the reason they were there wasn't the money or the fame; it was because they wanted to be.

Alan Whicker hasn't really been replaced – will David Attenborough be replaced?

Attenborough is famed for doing his many *Life of...* series, which were mega-productions which took four or five years to film. They employed the best cameramen, they had the best resources in terms of technology, and I'm not sure that there's that market any more – perhaps there won't be more of that type of programme made. Attenborough was able to make his programmes when the Americans wanted an English presenter, because they thought the accent meant authority, and at the moment they're into American presenters, which is perfectly understandable. So without that market you can't have the Attenborough figure.

Where will you be in ten years' time?

If I'm still alive – and I am one of those people who gets up every day and thinks, 'This could be it.' I still count full moons and I still stop for rainbows, because life is ephemeral and one of our crimes is that we forget that and think it goes on and on. So it's about cherishing the moment. I don't look into the past or the future; I'm very much in the present.

Would you like to be primarily a photographer?

I don't know. I very much go with the flow. Ultimately I hope to be working, but if I end my life sweeping the streets, as long as I'm being paid for it I'll be happy.

If I could introduce you to several people you've never met – alive or dead – who would they be?

George Armstrong Custer, as I'd love to know what happened in those final moments at Little Big Horn. Nelson, to find out if he is as we have depicted him. I'd also like to meet Audrey Hepburn. To meet her would be as good as seeing a male Bullfinch being eaten by a male Sparrowhawk.

What about within the realms of wildlife and natural history?

I was lucky enough to speak to David Attenborough a couple of weeks ago. It's always a pleasure when we meet up. I would like to have met Alan Whicker so I could ask him for a few tips, as I'm always keen to learn. The more I learn about Darwin, the more confused a figure I think he was, and there are certain aspects of him that I can't reconcile. I don't really yearn to have a conversation with other natural history greats. I meet a lot of naturalists and they are really the unsung heroes – such as Ron Hoblyn, Ian Newton and Sir John Lister-Kaye. They are not of high public renown but I have tremendous respect for them.

Where is the best place you've ever been birdwatching?

Hortobágy Fish Ponds in Hungary, when the cranes come to roost in October.

Where is the worst place you've ever been birdwatching?

Unquestionably Indonesia, where they kill everything.

If you could go birding to one more place in your life that you've never been to, where would that be?

Papua New Guinea, to see the birds of paradise.

What is your favourite bird group?

Raptors or ducks.

What is your most wanted bird?

Philippine Eagle.

You're on a desert island. What's your favourite piece of music?

'Shout Above the Noise' by Penetration.

Favourite film?

Once Upon a Time in America.

Favourite TV show?

I don't have one.

Favourite non-bird book?

F. Scott Fitzgerald's *Tender is the Night.*

Favourite bird book?

The Handbook of the Birds of the World (HBW).

PHIL HOLLOM

*Philip Hollom was born in the 1910s. He was an author of the first really good field guide to the birds of Britain and Europe, with Roger Tory Peterson and Guy Mountfort (*A Field Guide to the Birds of Britain and Europe*), which was first published in 1954.*

INTERVIEWED BY KEITH BETTON

What are your first memories of birds?

I clearly remember being lifted up at the age of four to peer into the nest of a Song Thrush. The memory is as clear now as it was then, and I was struck by the beautiful mud-lined nest and the bright blue of the eggs that it contained. As a boy I was fascinated by birds and I used to catch them using a garden sieve held up by a twig and a piece of string. Of course, that would be unthinkable today, but in the 1920s it was the only way that I was able to handle birds.

Were you encouraged in your birdwatching by your teachers?

I was given huge encouragement by my schoolmasters. Between the ages of ten and fourteen I attended Heddon Court School in Cockfosters. The headmaster recognised my interest and allowed me to wander around the agricultural land away from the school fields. I was the only boy allowed to do this, and I remember finding a Nightingale's nest and also a Hawfinch that had got caught in some garden netting. Back in those days there were so many birds in the area, and I recall seeing a Red-backed Shrike through the classroom window one July when I was sitting my exams!

When did your studies of the House Martins and Swallows begin?

In 1926 I moved to King's School in Bruton. This beautiful Somerset village had many nesting birds and I decided to carry out an intensive study over an area of four square miles. Harry Witherby was the man in charge of the national bird-ringing scheme and he provided me with rings, and in 1929 I ringed more than 250 Swallows in the summer.

You've inspired many people through your books. What were your favourite bird books when you were learning about ornithology?

I clearly remember my first bird book, and it was *Sketch-book of British Birds* by Richard Bowdler Sharpe. It was published in 1898 and I still have it today. Another favourite was *How Birds Live* by Max Nicholson, published in 1927. This book was particularly significant to me at the time, and I remember reading it under the sheets in my boarding school dormitory. For identification purposes, Thomas Coward's *The Birds of the British Isles and Their Eggs* was invaluable.

What happened to your birdwatching interest when you left school?

Well, I always thought university was a bit of a con, so I left school at seventeen and decided to get a job. At that time there were relatively few active birdwatchers. Harry Witherby remained a great influence in my early life, encouraging me to get involved in projects, and in 1930 he introduced me to Max Nicholson. I was just eighteen years old, and although Max was only twenty-six, he was already becoming a pioneer of bird study in Britain. Inspired by Max's own work on Grey Herons, I teamed up with ornithologist Tom Harrison and in 1931 we jointly organised a national survey of the Great Crested Grebe. There was a huge amount of work involved, when you think that all communication was via mail, and Tom handled nearly 5,000 pieces of correspondence.

What kind of help did you get in carrying out the census?

I was very lucky in knowing Marquis Hachisouka, who was a member of the Japanese Royal Family. Knowing that I had to carry out surveys of many pieces of water around the London area, he kindly provided me with an aeroplane, and together we flew around, plotting the newly created gravel pits that didn't appear on any of the maps at that time.

I know you moved to Surrey at this time, so where did you carry out your birdwatching?

I spent many happy hours at Brooklands sewage farm, not far from Weybridge. The highlight was finding an Avocet there in June 1932. I also spent a lot of time at Chobham Common, and Harry Witherby bought a house there, so there were always plenty of birdwatchers with him at the weekend.

How did the Second World War affect your birdwatching?

I joined the Royal Air Force in 1940 and travelled to Pensacola in Florida to learn how to fly. I spent eight months there and I often used to see flocks of birds when I was in the cockpit. I remember flying close to a flock of vultures on my first solo flight in Detroit, and thinking, 'If they can land safely, then I hope I can as well!' From 1942 to 1946 I was based in various places, and I was honoured to be flying VIPs on special visits. These included the Archbishop of York and the Anglo-American Committee on Palestine, who I spent six weeks transporting around the Middle East and Europe in a Dakota.

To what extent were these forays into Europe and the Middle East responsible for your decision to write a field guide?

Totally responsible. I used to arrive in a foreign country and find there was no book easily available to help me work out what the local birds were. It seemed obvious to me that we needed a European field guide. As a start, Bernard Tucker, editor of *British Birds*, thought it would be a good idea to reduce the five volumes of Witherby's *Handbook* into one concise book. Sadly, Bernard died shortly afterwards, but this was to be my first book, and *The Popular Handbook of British Birds* appeared in 1952 and remained in print until relatively recently. However, I was frustrated that we still needed a European field guide, and by chance Guy Mountfort, founder of the World Wildlife Fund (WWF), had exactly the same idea. Over lunch he told me that he had met Roger Tory Peterson, who was happy to do the illustrations, and he invited me to become the third member of the team. The result – in 1954 – was the arrival of *A Field Guide to the Birds of Britain and Europe*. I was chiefly responsible for producing the maps and working on the text. The book has now appeared in many languages and I am pleased to say you can still buy it.

You took part in the three famous expeditions organised by Guy Mountfort. What were they like?

I was honoured to be included on these expeditions. The first was to the Coto Doñana in south-west Spain in 1957, then in 1960 we went to Bulgaria and in 1963 to Jordan. If you've read the books, such as *Portrait of a Wilderness*, you'll know that Guy Mountfort was a master of organisational efficiency. Each member of the team had a role to play, and Eric Hosking was always the expedition photographer. Guy moved in lofty circles and had an amazing knack of persuading people to give us access to private places. Often he seemed to have a direct line to the Royal Family of the country concerned!

You've carried out a lot of private trips around Europe and the Middle East. Did you ever choose to travel with a tour operator?

Oh no, I always preferred to plan my own itinerary, and often my wife and I would set out from one place having nowhere to stay the next night. That would be unthinkable today, but we were so lucky. Wherever we went, people would offer us a room, and one man even lent us his house for a few days. Those were wonderful times. I really enjoyed travelling round Syria, Turkey, Iran, Tunisia and Morocco, to mention just a few. It was these excursions into the Middle East that motivated me to join forces with Richard Porter, Steen Christensen and Ian Willis to publish *Birds of the Middle East and North Africa* in 1988.

What was it like being involved in *The Birds of the Western Palearctic (BWP)*?

This was another brainchild of Max Nicholson. Work started in the early 1970s and a team of us collaborated to produce the texts. It was an unbelievable tour de force, with the nine volumes appearing between 1977 and 1994. It was quite an achievement, particularly as Max celebrated his ninetieth birthday in the same year as the final volume appeared. All this work was done at a time before the internet had been developed, and in the early days most of us didn't have computers. Now you can have the entire contents of *The Birds of the Western Palearctic*, with all the illustrations, on a single CD!

Looking back over the past hundred years, what changes have you have noticed?

There's been a huge increase in the number of birdwatchers and their access to information. When I started out, there were very few of us, and everything took a lot longer. Of course, there were many more birds in those days. I still keep notes of the birds in my Surrey garden, and in recent years we have had Mandarin Ducks nesting. When I started birdwatching they'd never been seen in Britain. While I've noticed the disappearance of many species, I've also seen how birds like the Jackdaw and Goldfinch have become much more common – and although House Martins and Swallows are less common in many areas, it was reassuring last year to see them in as good numbers as when I started my survey in Bruton in 1929.

STUART WINTER

Stuart Winter is a journalist and author. He was born in the 1950s.

INTERVIEWED BY MARK AVERY

Would you call yourself a 'birder'?

The very first article I ever wrote for a bird publication was on this very question. I sounded off in *British Birds* about the awful Americanisation of our culture and language, and said that the term 'birder' had infected birdwatching (note one word, no hyphen), especially when it was more appropriate for hunters than those who appreciated the beauty of birds and the thrill of watching them.

Thirty years on, I use the term 'birding' in most of the things I write. The term is punchy and perfect shorthand.

Am I a birder? As I say on my Twitter feed (@birderman): 'Stuart Winter is a birder, journalist and author, in that order. The journalism pays the bills – the birds make life worth living.'

I just wish life would give me time to do more …

How long have you been interested in birds?

My birding career began in the pram, literally! I was brought up in the East End of London and although it was a good few years after the Second World War there was still enough bomb-damaged waste ground – I hate that term, anything that has wildlife is not wasteful – to encourage wildlife to take up residence in the heart of the metropolis.

One day my mum was pushing me in the pram somewhere in the wilder reaches of Bethnal Green when I saw a bird high in the sky, seemingly unable to move. Mum's words had a massive impact on me as a two- or three-year-old.

'Ah, poor bird, it must be stuck in the wind,' she said with sadness in her voice.

The very thought that this poor creature was trapped in the sky and sentenced to a life of fluttering aimlessly had the same effect as being told bogeymen lived under the bed. I became terrified of being frozen in time and space whenever the wind blew. I never appreciated it at such a tender age, but I think I learned two important lessons that day: the fragility of birds (I think we must have been watching a Kestrel) but also that most important of birding credos – always look upwards.

How is your knowledge of plants, insects and other wildlife?

Having become a grandpa in recent years, I have recently learned the importance of looking down, as well as up. My grandson Benjamin is far more interested in picking up and playing with sticks, stones and mini-beasts than birds, but I am sure casual persuasion – emphasis on *persuasion* – will eventually see him join our number. He better had; I will need someone to drive me when I am in my seventies.

One thing about looking down at the ground at what Benjamin is picking up means I have been taking far greater interest in plants recently. I have always been impressed by the incredible knowledge of the 'botanisers' I have met in the field, especially on overseas birding tours. Not only do plant people know their stuff, but they have this great infectious desire to pass on their wisdom – birders have a lot to learn – with all manner of wonderful memory aids and mnemonics. One of my favourites goes: 'Sedges have edges, reeds are round, and grasses have rings that are easily found.'

I'm lucky that I live close to the Chiltern Hills, so I get a chance to put some of these little tips into practice.

Early influences?

I was born in London's East End in the 1950s and our family doctor had a wonderfully philanthropic outlook on life. She came from a background with links to the Bloomsbury Set and had a deep respect for nature. When we moved with my father's job from Bethnal Green to rural Bedfordshire it seems she took it upon herself to encourage my interest in nature, particularly birds, enrolling me in the Young Ornithologists' Club (YOC) and sending me tickets for an RSPB film show about Bobby Tulloch and Snowy Owls for my birthday. She also sent me a nest box to put up in the garden; the birds never took to it, but she incubated my developing interest into a full-blown hobby.

The other great influences in my life have been my school friends. School holidays were spent sneaking into Dunstable sewage farm or going on bus trips to Tring Reservoirs. Nearly half a century since we first went out with a shared pair of Navy field glasses, we still bird together. My school friend, John Lynch, and I recently took one of our mentors, Bedfordshire Natural History stalwart Don Green, to Minsmere to celebrate the fortieth anniversary of our first visit.

First pair of binoculars?

My first pair of bins was a pair of 10×50 Prinz from Dixons. I think they cost £19 – almost as much as my father earned in a week. I gave him £5 from the money I received for my twelfth birthday. He never asked for the rest.

What bins do you have now?

I love the Swarovski 10×32. They rarely leave my side.

First bird book?

The Observer's Book of Birds, what else? My first copy has long gone to birdwatching Valhalla – it was simply worn out – but I still have a recent edition on the bookshelf for posterity. The paintings by Archibald Thorburn are simply sublime. I recently bought one of the great man's pencil sketches. Perhaps one day I'll win the lottery and be able to afford a Thorburn oil on canvas.

How did you become a journalist?

Like most things in life, by chance. I always wanted to be a policeman but my shoulder-length hair and flirtations with Trotskyism at the tender age of sixteen were not the kind of things the Metropolitan Police wanted to see on a CV, or at an interview, in the 1970s. Guess they're prerequisites today… Back then, journalism did not quite have the appeal of the Sweeney. There was still the pursuit of truth and justice, but no opportunity to kick down doors and offer those immortal words: 'Put yer trousers on, you're nicked!'

Anyway, I wrote a letter to the local newspaper group but left it floundering on a cabinet in our sixth form common room for weeks before a teacher decided to post it. The result was an interview and a subsequent college bursary and, later, an apprenticeship on the *Hitchin Comet*. I got the front page story after a couple of weeks by jumping in a river and saving a baby. It's still the best story I have ever had.

Lots of people think that journalism is just about drinking lots and writing lies – are they wrong?

I am not exactly teetotal but I much prefer a cup of PG Tips to a pint. Most journalists do today. Long lunches and expense accounts are a distant memory.

Do we lie? On the whole, journalists, even after the Leveson Report and Scotland Yard's terrier-like investigations, are an honest bunch. I would wager – that's not a regular vice either – that far fewer journalists have appeared at the Old Bailey over the years than doctors, lawyers, policemen, clergymen or bankers. In fact, the only professional paragons I can think of are conservationists, and that's why I have long regarded myself as a birder first and a hack second!

How did you persuade the *Sunday Express* to let you write about birds?

My first weekly bird column appeared in the *Daily Star* in 1994 and was the result of a Fleet Street 'nipple war'. The *Sun, Sport* and *Daily Star* were all publishing Page Three photographs of young ladies in varying degrees of undress, and such was their popularity that there was readership pressure to use them to fill other pages, too. Apparently an order came down from the upper floors of the Grey Lubyanka (the old name of Express Newspapers' Blackfriars Bridge office, in honour of the KGB's Moscow headquarters) that only a certain number of female nipples could be seen in any edition. I seem to remember it was an odd number …

The *Daily Star*'s editor, the legendary Phil Walker, had a cunning plan. Phil was a keen birder and asked me if I would like to write a weekly column. I bit his hand off. After a few weeks, the *Daily Star*'s 'twitching page' became the talk of Fleet Street, not so much because of my exclusives on the arrival of rare birds – one reader actually found a Black-and-white Warbler in her garden – or what the RSPB was up to, but because people were speculating on the reason Phil had decided to give over a page of a 'red top' tabloid to the birding community.

'It was the only way I could get more tits in the paper,' he told one of the trade publications.

I later took the column with me to the *Sunday Express* when I was offered the role of Environment Editor by its editor, Martin Townsend. The 'Birdman' column has been running weekly for more than twelve years.

Does the press make a difference to our environment?

Environment stories rarely feature high up on the news schedule. I have worked on news desks and know that, when up against political exposés, celebrity scandals, dastardly crimes or 'village pump tittle-tattle', an environment story is rarely regarded by a cynical, careworn news editor as either sexy or, to use newspaper speak, a real belter.

Over the years, though, I have managed several front pages on issues relating to global warming and the impact of extreme weather. One of my favourites revealed how seal pups were drowning on Arctic pack ice that was melting because of the effects of climate change. I even managed to get the RSPB on the front page when the society decided to dedicate a number of new nature reserves to the Queen's Diamond Jubilee. 'Queen Saves Countryside' was the headline.

Are there other environmental journalists whose work you admire?

Geoffrey Lean gave my first assignment on eco-journalism: a piece on the decline of Kentish Plovers in Britain as a nesting bird. He has won countless awards, and although not too great of stature, is a true giant when it comes to making a case to protect the environment in a difficult world.

My other hero is the indomitable Michael McCarthy: crafter of words, winner of awards and champion of conservation. He has won medals from the RSPB, BTO and the Zoological Society of London (ZSL), and countless plaudits from his peers. I have had the good fortune to not only work alongside him but also bird with him on a foreign assignment. He is a great guy.

Do you think the environment gets fair coverage in the media, and are the environmental non-governmental organisations (NGOs) doing a good job to get their messages across?

There is no group of people more dedicated to any cause than the men and women who work in the press offices of conservation organisations. They produce an endless stream of good stories and commentary which, unfortunately, many sectors of the media ignore or overlook.

The small band of environment journalists do a good job getting stories into the respective publications but, as a whole, editors and news editors – and I was one for fifteen years – invariably take the weakest line of resistance and opt for a knee-jerk news agenda. The tabloids are obsessed with celebrity, and too many of the so-called quality broadsheets have a right-wing bias against anything green or eco-friendly.

I am lucky on the *Sunday Express* to have an editor who is supportive on issues such as animal welfare and the loss of the countryside. This is something that resonates with young people but, unfortunately, most people under the age of thirty get their news from the electronic media.

You've written two books about your experiences – *Tales of a Tabloid Twitcher* and *The Birdman Abroad* – how many more do you have in you?

Like all newspaper hacks, I have whole library of ideas for novels, most of them emanating from my days as a crime and terrorism correspondent. If I told you the plots, I'd have to kill you …

Seriously, working on a newspaper means turning out a non-stop flow of words every day, so I can only write – or think about writing – books during holiday periods, and that's when I do most of my birding.

That said, I am in discussions with a television producer and publisher on an idea which I am sure will delight birders everywhere. Watch this space …

You've met plenty of politicians involved in the natural environment through your work – who did you find the most impressive? And – if you'll tell us – the least impressive?

I can imagine the grimace of a career politician being summonsed to Number Ten after a general election victory, only to be told they will be working out of Nobel House – the home of the Defra. Perhaps the only job with a bigger black spot is Northern Ireland, but we'll get to Owen Paterson in a moment.

Over the years I have spoken to several environment secretaries. For some, it was just another political job and a chance to keep their seat in the Cabinet. Others wanted to make a difference. In truth, it has been the junior ministers who have impressed me most. Elliot Morley is a birder, and it showed when he was at Defra. He knew his stuff, and Defra civil servants knew that he knew his stuff. Sadly, his career was finished by his role in the expenses scandal, but I still believe he has a place in the world of eco-politics. One of my most favourite interviews was listening to Elliot describe how he had encouraged Tree Sparrows to nest in his garden. He was cock-a-hoop.

Richard Benyon might be on the other side of the political divide to both Elliot and me, but he has an inherent love of the countryside. He also knows how it works. One cold January day he showed me around his rather grand Berkshire estate and was thrilled every time we saw a bird, at one stage reversing his 4×4 *à la twitcher* when a Stonechat popped up on a hedge. It was a new bird for him on his estate, and he was really excited about ticking it off.

Prime Ministers seem to change their environment secretaries as often as their undies, so let's hope Owen Paterson's tenure is truncated at the earliest opportunity. I don't think my blood pressure will stand any more Badger culls. His tough, no-nonsense approach to politics may well have been suited for the Northern Ireland office, but I don't like the idea of having someone in charge of our natural world who openly supports Fox hunting.

I know that you are a vegetarian and a keen supporter of animal welfare organisations. Are you disappointed that more birders haven't followed that route?

Hands up time. First of all, I do eat a small amount of fish, as long as it comes from sustainable sources. I stopped eating meat several years ago and must admit that I feel all the better for having more pulses, vegetable protein, nuts and grains in my diet. My cholesterol levels are good, especially for someone who has had to watch their weight since giving up sport years ago. Writing about the horsemeat scandal and some of the appalling practices involved in putting animal flesh on the table vindicated my decision.

With the global population increasing at an unabated rate, one has to ask whether rearing animals for meat is a sustainable form of agriculture. Can we really justify pouring so much cereal into burping, flatulent, climate-change gas-producing livestock? Not only is this a terrible waste of primary foodstuffs – a huge percentage of cereal grown in the UK goes to become fodder for reared animals – but it is also a major contributor towards carbon dioxide emissions.

Healthy food and birding have never been great bedfellows. Cooked breakfasts, service station haute cuisine and fish and chips have been the fuel over decades of hard-core twitching and patch-watching. I look forward to the day when the twitching masses celebrate ticking off a first with tofu, tabbouleh and a cup of tea rather than a pie and a pint.

As far as animal welfare issues are concerned, there are some great birders out there who have become wonderful voices on issues such as blood sports and raptor persecution, particularly the irrepressible Charlie Moores and, of course, Bill Oddie. Get on to Twitter and see them flying the flag for greater animal welfare! I salute them.

Favourite film?

Zulu. My dad took me to see it when I was a nipper. I can still feel the goose bumps as the vast *assegai*-waving *impi* – arguably the finest foot soldiers in history – came over the horizon and charged towards Rorke's Drift.

Favourite TV programme?

Historically, *The Sweeney:* John Thaw, an actor's actor, playing a copper's copper. Recently, *Line of Duty,* arguably the most accurate police drama ever scripted. Tense and twisting.

Five pieces of music to take on a desert island?

'The Battle Hymn of the Republic' performed by the Mormon Tabernacle Choir; Pachelbel's Canon; 'Up the Junction' by Squeeze; 'Let's Stay Together' by Al Green; and the theme tune of *For A Few Dollars More*, performed by Ennio Morricone.

Favourite non-bird book?

Non-fiction: *In Cold Blood* by Truman Capote. It revolutionised journalism as well as book-writing; a true-life crime story that reads like a novel.

Fiction: *The Lord of the Rings* trilogy by J.R.R. Tolkien. I read it at the age of eighteen, then at thirty-six, and recently again. It gets better every time. Surprisingly, I have still not read the last couple of chapters. I get to the bit when the one true ring is destroyed (can there be a better denouement in literature?) and always decide I will leave the final pages for when I finally put my feet up …

Favourite bird?

What an unfair question! It's a bit like asking who is your favourite son or daughter. I am going to be greedy and opt for the Yellow Wagtail: brilliant colours, taxonomic nightmares and providers of countless identification challenges.

If I had to choose a whole genus then it would have to be the recently lumped *Setophagus* warblers of North America. Their colours span the spectrum, and trying to catch sight of them flicking through dense leaf cover is one of the most exciting birding challenges.

Favourite place to go birding – locally and worldwide?

My local patch goes under the atmospheric name of Sharpenhoe Clappers. It is a wonderful stretch of chalk escarpment at the northern end of the Chilterns which, on its day, is brilliant for Vis Mig (visible migration).

On a world stage, I would have difficulty choosing between Israel – that is, if I can go for a whole country – and Cape May on the eastern seaboard of the United States. Both are brilliant places for witnessing migration in all its glory.

Most recent holiday/overseas trip?

I have been lucky to do a fair bit of travelling with my work, covering major stories overseas. I even got to cover the football World Cup in Germany, fitting in looking for Grey-headed Woodpeckers and Marsh Warblers between covering the antics of Beckham and co.

My last trip was a prize in a travel writing competition: a week's cycling around Denmark. I think the second prize was two weeks' cycling ... Seriously, though, I went with my pal John Lynch and we had a brilliant time watching migrating raptors.

Do you have a birding/conservation hero? How about one still alive and one no longer with us?

I feel like an awful name-dropper, but there cannot be too many people who have birded with Sir Peter Scott, Roger Tory Peterson and Peter Grant. Sir Peter pointed out the first ever Gadwall I saw at Tring Reservoirs back in the late 1960s when he was attending a BTO conference and our school birding gang was on one of its Saturday adventures. He was with James Fisher. Double tick!

I was privileged to interview Roger Tory Peterson at his Connecticut home shortly before his death in 1996. He was in his eighties and had the ears and eyes of a Ninja, picking out birds in the grounds of his home as they came to pay homage to the great man. In the early 1980s I also spent some time with the late, great Peter Grant at Dungeness. He had a wonderful way of explaining the most complex bird identification issues in an easy-to-understand way. His premature death, while still in his forties, robbed us of a genius.

LEE EVANS

Lee Evans is probably the most famous twitcher in the UK.
He was born in the 1960s.

INTERVIEWED BY KEITH BETTON

How did you first get into birding?

My uncle Tony lent me his pair of 10×50 binoculars in 1968, when I was just eight years old, and took me with him on his visits to John Gaunt's Golf Club at Sandy, Bedfordshire. It was on this golf course that I saw my first ever Green Woodpecker, and I was so amazed at its bright colours and fascinated by its feeding actions that I spent a whole afternoon watching it. I used to spend hours at the RSPB's headquarters at The Lodge, Sandy, watching the nesting Sand Martins by the car park and the many woodland species that would visit the pond to drink and feed. Lesser Redpolls were common then, and Lesser Spotted Woodpeckers and Kingfishers were frequent visitors to the pond. Bonus birds came in the form of Great Grey Shrike, Stonechat and an unprecedented flock of Twite. And that's how I got the birdwatching bug and became committed (and addicted) to birding.

I used to find lots of dead birds at around this time during frequent cycling jaunts and took them all home, marvelling at their wonderful colours. My obsession with dead birds culminated in an interest in taxidermy and in those early years, I soon amassed a large collection of specimens, ranging from Atlantic Canada Goose, Common Sandpiper and Wigeon to Great Spotted Woodpecker, Bullfinch, Red-backed Shrike and Common Redstart. In 1973 I met a guy called John White, who had an old Vauxhall Victor, and we discovered twitching, and that's when I became obsessed with it and never looked back …

What about your time at school?

My education was at Denbigh High School in Luton. From the age of twelve, I converted a group of my friends into keen birdwatchers and we would spend the lunch hour visiting Wardown Park or checking the grounds of the school.

Lesser Spotted Woodpecker was high on the menu in those days, as well as Siskin and Bullfinch. The rarest birds I ever saw from within the school grounds were Russian White-fronted Goose and Black Redstart.

Tring Reservoirs became my special place, and provided me with my first real taste of 'rare birds', with a juvenile Pectoral Sandpiper in September 1969. My uncle Tony, who had a passing interest in birds, also took me to see the Black-crowned Night Heron at Lemsford Springs in 1970. The highlight of each week soon became the Saturday outing to Tring, and I would wait with six other keen youngsters for the 61 bus outside the Co-op in George Street, Luton. For just 54p return, we would be taken 42 miles in total, being dropped off at the Grand Union Canal at Bulbourne, and having nearly seven hours at the reservoirs.

Those early visits to Tring provided me with my first keen car-load – Matt Andrews, Phil Rhodes, Paul Fuller and Dirk Nelson. It was also where I met Derek and Jan Toomer and where the spark to form the South Bedfordshire RSPB Group [now the Luton and South Bedfordshire group] was ignited. Between 1973 and 1978, the Group was my life and obsession, and us five lads found friendship in John White. He was dead keen, and I managed to cajole him into travelling to the coast three out of every four weekends. Mark Simmonds, Paul Anness, Adrian Bonds and Meharben Singh also played a significant part in my life then too, often accompanying me on cycle rides or outings.

My first real interest in rarities started in 1974, when several of us would cycle to Norfolk and Suffolk for species such as Tawny Pipit, Black-eared Wheatear and Lesser Grey Shrike. By 1976, the highlight of my year was staying with Brian Mellow and Pete Maker in west Cornwall, courtesy of the YOC.

What happened when you left school?

I left school in July 1976 with seven O-Levels, and went on to sixth form college, where I took five further O-Levels and four A-Levels. Technical drawing, art and biology were my forte, and with these qualifications in-house I was accepted for an apprenticeship with Vauxhall Motors in August 1978. I then took a diploma in Mechanical Engineering and later graduated from Hatfield Polytechnic with a degree. My initial career at General Motors was fairly mind-provoking and varied and, after 'fledging', I joined the design team in AJ Block as a design stylist. The work was intense and for most of the time I was at the Millbrook test track in Bedfordshire, where I spent hours studying the likes of Lady Amherst's Pheasant, European Turtle Dove and

Long-eared and Short-eared Owls. I was known affectionately as 'the Bird Man' by the other employees, and it was a standing joke that I was either birding or sleeping on the job.

What was your first pair of binoculars? Your first telescope?

They were Swift Audubon 8.5×44s, which I used throughout my teens in the late 1970s. It took many years before I upgraded to a 'scope, and then only when I came across Optolyth in Luton one afternoon. I met the guy who was setting up this new business and, because he knew I was obsessed with birding, he sponsored me and kindly provided me with the best 'scope he had at that time, in the hope that I would promote it for him. As a result, I helped shift an awful lot of units of the classic Optolyth 30×75 telescope in those early days.

When did you join the YOC?

I joined the YOC in 1972, where I quickly rose through the ranks of the local branch to become Trips Organiser. Essential on every annual itinerary was an excursion to west Cornwall for a week, centred around the youth hostel at Cot Valley, St Just, with Roger Butts, Brian Mellow and Pete Maker being the designated leaders. They were the top Cornish birders of their time, and are still at the forefront now. I very quickly developed a good rapport with Brian and Pete, and enjoyed listening to their tales of American Kestrel and other 'old' Cornish 'rares' – the Bodmin Moor Kestrel being a bugbear of mine, me still being at school and dipping on the first available weekend. The bird was there for 16 days and seen by over 300 people – but not me! We did see Common Redstart, which was a new bird for me at the time, but we didn't get the American Kestrel – that's one of Brian's blockers on me to this day.

What about your early bird books?

I had a book called *Spotting Birds*, then a collection of books written by James Fisher, then Watson and Campbell's *Oxford Book of Birds*, the Audubon guides to both birds and mammals, and eventually became infatuated with the *Hamlyn Guide: Birds of Britain and Europe*. My grandmother bought me the *AA Reader's Digest Book of Birds* as a Christmas present in 1972 – the one with the Tawny Owl on the cover – a book I immediately fell in love with. This tome had all of the common bird species at the front, with the rarer/scarcer ones at the back, and I used to fantasise about how I could get to see these latter ones – and pencil-ticked each one whenever I did. At that young age I set myself the task of seeing all the birds at the back of the book, as did the eight other youngsters I cajoled into the hobby, and we had a little competition going between us. Instead of doing our school work we would regularly catch the 61 bus to Bulbourne and spend the day at Tring – oh, those were the days! Because I was so obsessed with birds, everything else took a back seat, much to my parents' despair.

Were you good at keeping notes?

I began collating field notes from my earliest outings in 1968 – most noteworthy from those early days being the aforementioned juvenile Pectoral Sandpiper at Startop's End Reservoir, Tring. By far the rarest bird I discovered in those early visits was a Long-billed Dowitcher on the mud at Startop's End that had relocated from Staines Reservoirs, as well as a pair of Kentish Plovers on Wilstone – a species that has not appeared locally since – and Egyptian Geese, a mega-rare then. From a very early age, I was obsessed about taking notes as much as I was about actually seeing those birds, and kept a meticulous record of them in foolscap.

So how did you get into twitching?

My twitching career began with a vengeance in 1977, following two enjoyable holidays on the Scillies in 1975 and 1976. I had met fellow twitcher John White, who just happened to have a car – a trusted Vauxhall Victor SL – who drove me and the others to the latest 'rare' each weekend. By 1978 I was totally hooked, and chasing every rarity that I could. Unfortunately, that year I had a very serious road accident, losing my eye and spending many months in hospital, including Christmas 1978, when most birders were out enjoying the first Greater Sand Plover for the UK – at Pagham Harbour in West Sussex. My best birding pals at around that time were Dave Holman, Graham Ekins, Chris Harris, Dave Rosair, Arthur Livett and Barry Nightingale, and I was very grateful for their support through that difficult time. Following recovery from the accident, I then became utterly obsessed with listing and, from 1979 onwards, I concentrated on year-listing and building up my British life list. It was also the year that Matt Andrews and I spent all summer in Europe, traversing Austria, Germany, Greece and Turkey in our quest for European species. We arrived home in late August with over 320 species, including a flock of Slender-billed Curlews from Porto Lagos.

When did you first go overseas birding?

I learned to drive when I was seventeen, and my first car was a tiny Mini my dad bought me. I then upgraded to a red Vauxhall Chevette Estate, acquired from my place of work. I was at college then, and when we broke up on 21 July,

Matt Andrews and I went on a seven-week trip across Europe trying to see as many birds as possible, all going pear-shaped near Porto Lago in Greece when some character decided to do a U-turn in front of me as he departed a garage forecourt – the two of us colliding head-on. My Chevette was badly damaged at the front as a result of the impact. We had no headlights, a punctured radiator and a crumpled bonnet, and had to drive all the way from southern Greece to Calais with no lights. We never got to see our target bird either – Yelkouan Shearwater – and spent a whole day and a half at a garage getting the car welded back together. It did allow us plenty of time to study eastern Rufous Bushchat and Olive Tree Warbler though!

In the late 1970s, who did you really look up to in birding?

Throughout my progression into twitching, Norwich birder Dave Holman was my hero. He was a major cog in the grapevine wheel back then and for many years, I would phone him almost daily. He was an outstanding birder – and a brilliant teacher – and I would pester him non-stop for information on 'rares' or just identification matters. Richard Richardson was high up there too – and I revelled in spending hours on Cley's East Bank watching Turtle Doves and Linnets stream west in their thousands every April with him. I was in my element. Peter Grant was also one of my great birding heroes, often taking me under his wing, post-1979, to explain difficult IDs such as Ortolan Bunting and Yellow-browed Warbler. And, of course, there was my lifelong pal Keith Vinicombe, who was instrumental in opening up my mind to birding trivia and the like during our many hours together in the Turks Head on St Agnes each October.

As I grew older and more experienced, other observers came into my life. Killian Mullarney instantly standing out as one of those gifted with exceptional skills, as well as Dick Forsman, Richard Millington and Annika Forsten, international observers met during my repeated visits to Israel post-1982. Paul Holt impressed me a lot with his capacity to seek out rare buntings and pipits, predominantly on call, while Chris Heard and Grahame Walbridge were the 'real deal' – Chris's eyesight and hearing being second to none. Steve Whitehouse and John Belsey must get a mention also, as well as Ray Turley, Clive Byers, Craig Robson and Jon Eames: all were outstanding in the field and very trusted observers.

Throughout the 1980s and 1990s my life was pretty much non-stop birding, so much so that from 1987 I decided to take it up professionally. I took early redundancy from my design stylist job with Vauxhall Motors and co-created the Bird Information Service with Richard Millington, Steve

Gantlett and Roy Robinson. During the first five years, the business went from strength to strength, the monthly journal *Twitching* almost immediately morphing into *Birding World*. The accompanying Birdline phone information service proved to be a real winner and, for over five years, we virtually had no competition. Nothing this good lasts for long, however, and by 1993, revenues were falling fast and it became inevitable to dissolve the partnership that had proved so successful for seven years. I then concentrated on my own publications and the UK400 Club (established in 1978), and for many years successfully produced a catalogue of rarity-orientated books and a bi-monthly magazine, *Rare Birds*. A catastrophic mistake in 1998, however, saw the business and empire come tumbling down as my cherished Vauxhall Cavalier GSi was stolen as I offloaded magazines into Post Office sacks one morning. I had lost over £30,000 in one stroke, and had no option other than to declare bankruptcy. Everything I had ever owned of value was literally in that car, including a number of brand new 'scopes and binoculars I was reviewing.

How did you meet Carmel?

I was resident DJ at Cinderella Rockafella's in Dunstable town centre from 1978 through to 1982, and met Carmel one Friday night as I gigged there. I took an instant liking to her, and our relationship just blossomed and sparkled from that evening on. From April 1987 through November 2000, she was always to be found at my side, enjoying birding in the fast lane. Sadly, though, when she discovered she had cancer later that year, her interest faltered and it was never rekindled. Her life was then taken over by looking after her increasingly frail mother in 2008, nursing her until she died in September 2013.

What do you think about the future of twitching and birding?

Twitching has changed so much in the last ten years – for the worst in many ways, I am afraid. We now affectionately talk about the 'Golden Years of Scilly': the islands as we came to know them 'birdingwise' dying out in Year 2000. The new millennium saw the advent of a new birding world: the beginning of a period when technology took over the hobby and perhaps ruined it for ever. Seemingly, there was no turning back.

October 1999 had been an outstanding year on Scilly, with some 4,000 birders reaping the benefits of some outstanding vagrants, but birders are fickle – the pager was now taking over from the phone information services and life lists were coming in at an average of 480. Unexpectedly, and taking me by great surprise, less than 500 birdwatchers appeared on Scilly the following October and, from that year on, the numbers just never recovered. It left a big lump in my throat. I was gutted. Scilly had been the be-all and end-all of twitching for decades. What had happened?

Although I continued to visit each autumn on day trips, I never did return to stay – Carmel's untimely illness putting paid to that. And now, just a distant and fading memory – Ray Turley, of course, passing away (Ray and Janet had spent many years with us at the timeshare in Godolphin Flats). And 2013 proved to be the most dismal year yet, with a lacklustre set of 'show-stoppers' and just 200 birders visiting in all (although I did just make it down, securing my first ever Caspian Stonechat on St Agnes in November). A terrible shame particularly as the islands had been the life and soul of birding for several generations. Scilly throughout the 1980s had spawned endless birding relationships and had pioneered countless numbers of international adventures, as well as forwarding the knowledge in rarity-finding and identification. Towards the end of the era, there had been an influx of young blood, with the likes of Chris Batty, Andy Holden, Alan Clewes, Andy Clifton, Richard Bonser and Stuart Piner appearing, replacing the likes of Ian Barnard, the Rugby Boys and the Whitcombs of the decade before. So there had been just under forty years of Vintage Scilly – from the ranks of Ron Johns, Ian Wallace and Edwin Welland from 1964 all the way through to 1999, three generations of the maddest, keenest twitchers in the land, and some of the best field ornithologists that this country has ever produced.

From a personal point of view, the advent of the internet has changed birding beyond recognition. It is now a place of the 'virtual birder': you can log on to websites 24/7 and get all of the information you ever need and not even have to spend a penny. Effectively, you have an ever-increasing band of people 'watching' birds on the internet. Many do no physical birding whatsoever. It's like fantasy football, where nobody bothers to go to watch a game – it is just fantasy birding. Furthermore, you can log on to the internet at any time of the day and find many of these characters arguing cat and mouse about the latest 'rares', seemingly not having a clue what they are talking about. They are not field birders at all, and have no experience. They have not served any form of birding apprenticeship, and seem to think that the be-all and end-all of birding is argued out on Birdforum (www.birdforum.net/). In the field we have a new breed of 'birdwatcher': the photographer with no binoculars and no field craft, an increasing sight at everyday twitches.

Let's go forward twenty years – what can we expect?

I see very few of the 'Old School' now at twitches – the vast majority have retired. A lot became disillusioned by the sport and concentrated their efforts on world birding. Some would say that they grew up. I think more and more twitchers will fall by the wayside as the potential for adding new species to their life list dries up. The decrease in the population of many species has had an effect on the hobby too, while the cost of getting to birds continues to increase year on year. Because of the high price of fuel it now costs an average of £60 just to twitch bread-and-butter year-ticks such as Ortolan Bunting, Barred Warbler and Red-breasted Flycatcher, while species such as Aquatic Warbler are now virtually impossible to find. For example, 2013 was the first year I had not seen a Golden Oriole in Britain for over thirty years, and Turtle Dove numbers were in such decline that I saw just eight during the year.

Although there are fewer proper twitchers today, there are a lot more people watching birds, aren't there?

Yes – there are now well over a million members of the RSPB, but unless there's a really rare bird like a Baillon's Crake in London, you just don't see that many people at the average twitch. I travel to see an array of rarities every week and, at many, you will just see a handful of watchers there. You turn up at a Black Duck twitch – and you may be the only one there! And, for example, when I go to see my local Great White Egret, which has spent three consecutive winters in the Chess River Valley near to where I live, there are many days when nobody else even looks for it. People see one, and then it's of no consequence to them, it seems!

What do you think about the general attitude of twitchers?

In general, it's appalling. Youngsters today have little respect for their elders. They go wading out into marshes where there are Baillon's Crakes breeding, they go stomping out into people's gardens where there are Baltimore Orioles feeding at the bird table, and they fail to take notice of local restrictions on access.

Few have any real interest in any bird's welfare, and very few of them are members of any organisation, bird club, RSPB, BTO or whatever. I just don't understand it – they don't understand the rationale behind the ethics of birdwatching. There seems very little evidence of any apprenticeships served, many coming straight into twitching, often seeing the rarer species before the commoner ones. Hence why many have not seen a Lesser Spotted Woodpecker – a bird you actually have to work hard for in order to see.

I was looking through the British Birds Rarities Committee (BBRC) reports through the 1980s and was looking for your name, which came up around 1985/86 with loads of birds you were involved in finding, but I haven't spotted it so much in recent years.

I lost interest in submitting bird records around the mid-1980s, as I found the politics of birding incredibly frustrating and self-defeating. That's why I set up my own organisation – it was just a shame I went bankrupt before I could fulfil all of my publishing goals.

What do you see as your birding achievements?

In terms of birding, recording a record 386 species of bird in Britain and Ireland in 1996; a whopping 682 species in the Western Palearctic in one year; 209 species in one week in Britain in May 2013; and 221 species in January. Also the fact that I have maintained a top-ten position in over 30 individual counties in the UK. I also languish in sixth position in the life list rankings, recording no less than 578 British Birding Association (BBA)/UK400 Club species in Britain and Ireland.

I was involved in editing *Twitching* and *Rare Birds* for many years, as well as *Rare Birds Monthly*, but I now concentrate producing monthly and annual summaries for *Birdwatching* magazine and to updating Twitter for my thousands of followers. I have written a total of thirty-seven books on ornithology, including the highly acclaimed *Status of Rare Birds in Britain 1800–1990* and four different versions of *The Ultimate Site Guide to Scarcer British Birds. Rare and Scarce Migrants of...* volumes have been published for Scilly and Norfolk, with Buckinghamshire and Hertfordshire currently in the pipeline. I have also written numerous e-books and e-identification papers too, including works on redpoll identification, White and Pied Wagtails, ageing Waxwings, Eastern Black Redstarts and Hume's Leaf Warblers.

Where's the best place you've been birding?

My lifelong favourite has always been Israel, particularly Eilat for the spring migration and the Hula Reserve for the diversity and sheer numbers of birds. Turkey is also high on my favourites list, particularly in the east, and of course the mountains of Georgia, at Kazbegi. Following that, northern and southern India excel – like watching a feeding party of Giant Hornbills crash overhead through the canopy. You can sit in the woods on your own and suddenly a flock of birds will come through with fifty-five species in one flock. I love North American birding too, and find that birders are far more appreciative of my contributions there.

Where would you choose to go if you could go somewhere you've never been before?

Probably Japan in winter for the twelve overwintering species of crane and Blakiston's Fish Owl – but I can't justify the £4,500 price tag.

Favourite bird?

As always and forever, Wallcreeper and Snowy Owl.

What sort of music do you like?

My all-time favourite singer is George Michael; he is an exceptional songwriter, with 'Careless Whisper' being my favourite song. I also love Soft Cell and Marc Almond, Erasure and Bronski Beat, the Pet Shop Boys and the Eagles. My most-played song is 'Comfortably Numb' by Pink Floyd, narrowly pipping 'Hotel California' and 'Say Hello, Wave Goodbye'.

Why do you often compare your life to that of George Michael?

Our lives have been very similar. He is top of the game in his penchant as I am but just doesn't get the credit he deserves. He is always making the papers with his notoriety, but very few people know the 'real' George. He polarises people – they either love or loathe him; there is no middle ground. He is a genius songwriter but suffers greatly from depression. I see parallels with our lives – we are always talked about and despised by those that only read the headlines.

What about films?

I watch very little TV other than *Question Time,* and the twenty-four-hour news channels, and have no penchant for films.

Books?

Sad, really, but my only reading is bird-related or pop music-related.

If you could take a bird book to a desert island, which would it be?

The European Bird Census Committee Atlas of European Birds.

STEVE GANTLETT

Steve Gantlett was born in the 1950s. He has seen more bird species in Britain than anyone else, and until recently produced the magazine Birding World.

INTERVIEWED BY KEITH BETTON

Where did it all start for you in terms of birds?

In 1961, when I was seven years old. A bird came to the garden bird table where I lived in Ashtead in Surrey, and I found a bird book in the house and worked out that it was a Meadow Pipit. It grew slowly from then, beginning with just a general interest in garden birds, and the YOC just happened to start not long after that. By that time, my mother knew I was interested in birds and encouraged me to join. I received the very first YOC magazine – which, annoyingly, I sold much later.

Did you write in to *Bird Life* magazine?

I did. I tried writing one or two articles.

Were you involved in any local YOC groups?

The YOC leaders used to run courses and holidays, several of which I went on. I'm not sure they do them any more. In early 1968 we moved to North Devon, but there were no other birdwatchers in my school, so I mostly birded alone until we moved to Hampshire in the summer of 1970.

Did you tell people at school of your interest?

Yes, I think they knew but I don't think they were interested. There was a YOC leader who a couple of times took me out on the back of his motorbike to see seabirds on the Exmoor coast. I think he was the honorary warden of the RSPB Chapel Wood Reserve.

Did you submit records to the Devon Bird Report?

Yes, a few. We were in north Devon from 1968 then moved to Aldershot in July 1970. My mother was from London and found Devon too quiet! By then

I was much more into birds. In my last year in Devon I was able to drive a motor scooter and used to spend all my time going to Tamar Lakes, which became my patch, along with Northam Burrows near Bideford. I did wildflower surveys on Northam Burrows.

I started at Farnborough Grammar School in autumn 1970 and Frensham Ponds became my local patch. In the Surrey Bird Report 1970/71, when they used to publish initials next to every record, I had initials on virtually every page because I was doing Frensham every day after school. A Spotted Crake was the best bird I saw, and watching nesting Hobbies. I think I was still wearing my YOC badge at that stage. At my new school I met Richard Millington, who was already at the school in the same class. I remember Richard coming up to me, seeing my YOC badge and asking what my list was. I hadn't a clue, so made up a figure (about 300) which was far too high, so he was very impressed! And we've been birding buddies ever since.

Did you do Fleet Pond as well?

Richard actually lived in Fleet so the Pond was about 100 yards' walk for him. There was a bit of rivalry between us as to what we could find. John Clark was at the same school, though a year above, with Richard's brother Spike. We also had a keen zoology master who was a birder: Ewart Jones, one of the famous Portsmouth Group in the 1950s. He took us up to Norfolk in a transit van a couple of October half-terms, which cemented my love for Norfolk – I'd got a lot more savvy on birds and realised that Norfolk was the place to be.

What was your first pair of binoculars?

My dad's old pair. A French make, I think. They weren't too bad.

And telescope?

A very cheap black zoom thing with an objective of not much more than an inch. For about a year I used only this telescope as my binoculars were stolen.

First good 'scope?

Swift Telemaster. I remember using it on a tripod on a Short-toed Lark twitch at Teesside, and I was the only person with a tripod, and because of the distance and terrain no one could see the bird but me, so absolutely everyone (about fifteen people) had to queue up to look through my scope.

What did you get next?

After my binoculars were stolen, the zoology master managed to persuade the school to buy a pair which somehow became mine – I'm not sure how that happened! They were a pair of cheap Russian Zoltz, quite widely available at the time, but not really good enough. In 1971 got a Saturday job, which cut into my birding time, in Marks & Spencer in Aldershot and as soon as I could, I got a pair of Leitz, paying £3 a week for two years. Leitz weren't the big brand they are today (Leica); everyone had Zeiss Dialyt. The Leitz were more expensive than the Dialyts, but better so I got them anyway.

Did you stick with Leitz/Leica?

I've always been a Leica man; I'm not one of those people who change every year. I still think they are the best, but there is strong competition out there nowadays, of course.

And what about 'scopes?

Those have varied over the years. At one stage I had one of the brass draw tube 'scopes, and then upgraded to the Telemaster at about £60. I used that for many years, then Nikon gave me a 'scope: they decided they'd give some 'famous' birders a 'scope to try and get their name out there. They gave one to Ron Johns and half a dozen other people.

What about your first bird books?

It was the Christmas after I'd spotted the Meadow Pipit in the garden. It was a dreadful book called *Bird Spotting*, which I sold for 50p when I was incredibly hard up at one stage. Before that I remember my sister being given *The Observer's Book of Birds* by an aunt; I was given *The Observer's Book of Pond Life*, and I was cheesed off that she'd got the bird book as she had no interest in birds!

Can you remember any particular books from your teenage years?

My first field guide was the *Collins Pocket Guide to British Birds* by Richard Fitter and Richard Richardson. I had that for years before I got Peterson, Mountfort and Hollom. I went on a YOC holiday and found that most people were using Peterson, and shortly after that I got one myself, though I always preferred the Heinzel, Fitter and Parslow one, which was my bible for five years or so.

So, when you were in Devon, you had the YOC leader with the motorbike who was an influence – anybody else?

Before we moved to Devon I was at the Freemen's School in Ashtead and there was a very keen biology teacher – not a birder, but enthusiastic about natural history and biology. I said to him there wasn't a Natural History Society at the school, and could we start one. He was very enthusiastic so I started it, but about six months later we moved. In Devon, which was a very nice part of the world and where I went to a nice school, there was nobody like that in terms of staff or pupils. I eventually made contact with the guy with the motorbike, but I only went out birding with him a couple of times, so he was hardly a big 'influence'. I did most of it on my own by reading what happened to be in the YOC magazine.

What was your personality like at that time?

I've always been quite shy. Looking back, it surprised me that I actually went to the biology master and asked about starting a Natural History Society, but I felt so passionately about the subject.

What would you say was your first real year of twitching?

I met Richard Millington in September 1970 and he had a few contacts, including Alan Mitchell (of Collins Tree Field Guide fame; *Trees of Britain and Northern Europe*), who used to run adult education courses on birdwatching. Richard and I went along to those with a lady called Dorothy Herlihey, who was a very keen birder who lived near Richard. For ages I thought she was Richard's mother – he was always out and about birding with her as she lived right next to the Pond. She was a great influence.

Who did have the biggest list – you or Richard?

I suspect we were about the same. Spike wasn't into twitching – he just went to Fleet Pond and for some unfathomable reason (to me anyway) he got heavily into Canada Geese and seemingly little else and we drifted apart. I had transport, a motorbike initially – I was the only one who could drive, and Richard used to come out with me on the back of my Honda 50. We even took it up to Norfolk.

Going on through the later 1970s, who were your influences then?

Every first Sunday in the month, Alan Mitchell used to take a gang of Farnham/ Aldershot/Fleet area birdwatchers to Selsey Bill to do a dawn sea-watch, then to Pagham Harbour and that area. Most of the group were middle-aged; we were the only young people. At Selsey, one time, we met Gerry Price, who entertained us with stories of twitching and the Isles of Scilly, and he became our grapevine contact, telling birders what the latest news was. At Farnborough Grammar there were several slightly younger lads, Graham Stephenson being the main one. There was also Tim Doran, Barry Tuck, Dave Buckler and a couple of others. As soon as I was seventeen I learned to drive and got myself an old banger (paid for by that Marks & Spencer Saturday job) and we spent a lot of time down in Portland at weekends. It was all down to finance at that time, so we used to fill up the back of the car with the younger lads so they could chip in with the petrol money. We used to go to Portland for dawn, driving this old Austin A30 full of fourteen- and fifteen-year-old boys, and I remember being stopped by the police on more than one occasion. That car only lasted a year, and then I had an old Morris Minor van. Once, the suspension collapsed when it was full of lads coming back from a twitch in the middle of the night. Richard hitched off up the road and left us, but we got a local garage out and the owner took it, and us, back to his house and gave us a meal.

When did you move away from Aldershot?

I trained as an optician at Batemans in Basingstoke from 1975 to 1978, so I didn't have quite so much free time. I gave up doing Frensham quite as thoroughly, and was twitching at weekends. Then in July 1975 I finally left home. I fancied a transfer from Basingstoke to either Weymouth or Canterbury, which were good places from a birding point of view. I'd often spent weekends at Dungeness and Stodmarsh. I asked Batemans if there was a chance of moving to either, and I moved to Weymouth, where I quickly met David Fisher, who was at teacher training college in Weymouth, and became friends with him. Weymouth was a great place for birding; I used to spend every lunch hour at Radipole.

1975 was a great year for rarities, wasn't it?

That period was actually the time when I decided to get out of optics because I just hadn't got the time I needed for twitching. There was the Black-and-White Warbler, the Sapsucker and the Tanager on Scilly, and I couldn't go because I was working. Eventually I got down to Scilly but all three had gone, and I decided to change my job, which it took a while to do. Until August 1976 I stayed as an optician in Weymouth, where I met Sue, who is now my wife – she was the optician's receptionist. Then I finally completed my training and I had to move on from Weymouth. I'd become very friendly with Iain Robertson, the Portland Bird Observatory warden at the time, and done lots of birding with David Fisher.

I stayed in contact with Richard Millington, but didn't see much of him as he was at art college in Brighton and didn't have transport. Batemans had branches all over the south-east of England but I wasn't interested in any of them unless it was somewhere I could go birding. Anyway, a job came up in Thetford – not ideal, as it was too far from the coast, but at least it was in Norfolk, and the job came with a car which was very attractive. So I was an optician in Thetford from August 1976 to November 1981. It was OK, but I still realised that my twitching was suffering from not being able to go for birds during the week.

You obviously got to 400 sometime in the early 1980s?

I'm not that hung up on numbers, really, but it all depends on what you count and don't count. I do remember I was going to Fair Isle by that time and found what I considered to be my 400th – Yellow-breasted Bunting – which at that stage was annual on Fair Isle.

Before you reached 400, did you find there was a 'You've got to reach 400 to be taken seriously' attitude among other twitchers?

Not really, no, and it wasn't really a goal for me. I used to go to Cley from Thetford every weekend and David Fisher used to come up at that time – before the days of Nancy's Café everyone used to meet in the George. Dave Willis was one of the regulars and I remember him calling a car-load of young Midlands lads 'The Leicester Low-Listers' – some of them are well-known and well-respected ornithologists now!

Did Birdline follow shortly after that?

Birdline evolved mainly out of Nancy's Café and that era. When I lived in Aldershot, Richard and I were going up to Norfolk quite regularly, or to Portland or to Selsey if we had less time or petrol money. Incidentally, the first weekend I moved to Weymouth there was a Terek's Sandpiper at Cley so I went straight back up, but I missed it. During my second weekend as a Weymouth resident was the White-tailed Plover at Packington, so I didn't see much of Weymouth in my first two weeks!

Were you year-listing as well?

Yes, I did at that period, but not frantically, except for 1980 when Richard tagged along almost everywhere and did his book (*A Twitcher's Diary*). But it was me doing the year list. I went to Fair Isle that year and Richard didn't, and he managed to grip me off with the Sooty Tern in the Midlands because of that.

After a year or so at Weymouth/Portland I moved to Thetford, and then I visited Cley every weekend. You had to go there to get the news, basically – anyone who was anyone went to Norfolk in those days. Then Nancy's Café started quite gradually and people started asking if they could leave messages and it just snowballed, and Nancy's became the grapevine. Birders used to come and sit in the café and just answer the phone, but eventually it was just ridiculous, as so many people wanted information they couldn't get through, so something had to give – and that's how Birdline started in 1987. I think we first discussed it on Scilly, and Lee Evans and Roy Robinson were also involved. We realised that a fair amount of effort had to go into it, so it had to be funded somehow, so we started *Twitching* magazine at the same time to help pay for it. I'd been tinkering about with desktop publishing of checklists and I had one of the very first Amstrad computers. We made it such that if you subscribed to the magazine, you could get the Birdline number, and hence the news. But the number got handed round anyway, so this was not going to last long. I was doing most of the magazine work and Richard was doing most of the stuff on the answerphone (of which there was only one) and by pure coincidence the 0898 phone technology came along at the same time, and they approached Richard to see if he'd set up a rare bird information service.

Did you and Richard set the business up together?

It was very much a partnership: Richard, Lee Evans, Hazel Millington – who was going to do some of the office work – and I were all involved. We had the expertise, but Roy Robinson actually had an answerphone service with the Birdline name. We called ourselves Rare Bird Alert – subsequently pinched by the pager people – and we only called ourselves that because Roy had the Birdline name. We approached Roy and asked him to come in with us and all work together. He took on some of the magazine office work and eventually dropped out. Richard, Hazel, Roy and I were all in Norfolk and Lee was in Buckinghamshire, and this did not work well, so Lee dropped out. Roy then dropped out. Richard and I carried on and the magazine gradually grew and, after a year, became *Birding World*. Very quickly we had Peter Grant on board, which was a great boost to us and to the magazine, but sadly that lasted only a year or so as Peter died. Simon Harrap assisted for quite a while too.

In less than ten years, pagers arrived and we looked into that technology too but there were various other rival birdlines setting up at the same time. There was a demand to regionalise – and the only way for us to do it was to get the key people in the regions, which we managed to do very well. For example, we had Dave Holman in Norfolk, Chris Heard in the south-east and so on. We worked well as a team and they called themselves Birdline South-East, Birdline North-East, etc. That way, we were all branded under the umbrella of the magazine rather than a lot of birders squabbling with each other which we saw as being a dangerous possibility, with people not knowing where to phone any more for news. The Birdline groups still exist, with part-time operators.

What about international travel?

My first foreign trip was to the Camargue in 1976 with Graham Stephenson and Dave Buckler, two of the 'sacks of potatoes' from the school days – 'sacks of potatoes' being a Tony Smith-coined term for petrol-money-paying birders! Tony used to go round with a group known by other people as 'the String Band', which was himself, Lester Mulford, John East and one or two others, who were always down at Portland or in Norfolk.

We did the Camargue and the Alps, and it was wonderful. Next trip was the following year to India for two weeks with Jake Ward, Pete Milford and Kerry Harrison, to Delhi, Ranthambore and Jim Corbett National Park, which was a new experience for all of us and completely blew our minds. It was nowhere near enough time, of course, so that led to the resolve that something had to give. At this stage the economy was in recession; I'd bought a house in Thetford; I was earning a reasonable salary as an optician, but the mortgage payments had gone through the roof and seemed to be taking all my monthly salary, so I found myself working five and a half days a week paying for nothing but the house. Yet the value of the house had gone up enormously in a couple of years, so it dawned on me I could sell it, get a nice lump sum, get out of optics and go off round the world birding, which seemed like a far better way to spend your life!

The plan was to go round the world for a year with Dave Haslam, but it took me a year to sell the house. In the meantime, he'd got married, so he wasn't up for it. So, come autumn 1981, I resigned from the opticians. I'd spent the whole of the previous year planning the trip so I knew exactly where to go. Best thing I ever did. But I always wanted to spend as much time on Scilly as possible every October, so I spent the whole of October on Scilly and then took off round the world in November.

I started in Nepal and went across Asia to the Philippines and Australia. At that time there were still some wild Californian Condors left, and August was the recommended time to be there so I finished up there. Also there I did a pelagic with Debbie Shearwater when we saw a Humpback Whale; the first she'd seen. I had a great couple of weeks there with Paul Lehman and John Dunn.

I wanted to get back to England for the autumn, and I'd made up my mind I could do the 'difficult' places first and save North America for my old age. I came back to England in late August 1982 – it had been a fabulous year for rarities back home, I'd missed loads, such as the Marmora's Warbler and Hudsonian Godwit. Gradually, over the years, I've got them all back except the White-crowned Black Wheatear. That good year carried on through the autumn, and the week after I got back there was the Long-toed Stint on Teesside and the Little Whimbrel in south Wales, and Scilly was really good as well.

I hadn't quite run out of money at this stage and on Scilly I met Richard Fairbank, who said he was doing a trip to Venezuela for two months – the Venezuelan field guide hadn't been out very long and it was the 'in' place to go. So we went together in November 1982.

How much did the around-the-world trip cost?

Probably about £5,000. The house I couldn't afford in Thetford cost me £10,800 and I sold it for £16,000, which at the time seemed like a fabulous profit. When I came back I could still get a job as an optician if I wanted to, but I'd have to start again in terms of a house.

Were there any points during the world trip when you thought better of it?

No, it went far better than I ever thought it would. I was worried about loneliness but in fact it was a good way to do it because you have to meet people, and you get to meet people, much more than if you're with another person or in a group, when you just tend to talk to each other.

The last straw with regard to twitching when I was an optician was the Magnolia Warbler on St Agnes in September 1981. In those days I used to work Saturday mornings in Thetford and in compensation I used to get Tuesdays off. I remember the Magnolia Warbler was on a Sunday and everybody twitched it on the Monday, so rather than try and pull another sickie I just had to bite the bullet and wait till Tuesday, but it had gone then and I dipped and that was it for me. I thought, I've got to get out of being an optician!

How many species did you see on your world trip?

With Venezuela, which took it over the 12 months, I think I was one of the first – if not *the* first – to see 2,000 in just over a year, which was some sort of record.

It didn't, of course, cure me of birding; quite the opposite, really. I think I'd told my parents that when I got back I'd do the optician stuff again, but they could see that it wasn't really the right thing for me. Mother probably worried about me going round the world, as mothers do, but not seriously.

So I came back from the world trip, and I don't know if I was kidding myself, or them, with this idea of going back into optics. This was about the time when the first computer came out, from Amstrad, and I wrote up some of my trips in a booklet called *Bird Trips* and sold them to birders. Also, I started to produce country checklists and after a while got into that with David Fisher, as he was now leading holidays for Sunbird birdwatching tours. I also produced a book called *Where to Watch Birds in Norfolk* which I sold quite a lot of, and that's how I got into desktop publishing and the idea of *Twitching* cum *Birding World* came along.

I nearly got a job with Lawrence Holloway as a regular tour leader with Ornitholidays in Bognor because my parents had retired down there and he was based there, but it was never really going to work – I wanted to live in Norfolk, not Bognor. It was British birding that I was really keen on, so I wanted to get back to Norfolk. Neil Bostock had a house near King's Lynn and he'd been renting a room to Robin Chittenden who'd moved out, and I took his place. There was a government scheme called Enterprise Allowance – you got something like £40 a month to start up a business, so I went on to that for a year trying to sell the bird booklets and doing quite a few slide shows for bird clubs, barely making a living but being able to go birding a lot of the time. Then everyone realised that the best rare birds could be seen on the Scilly Isles in the autumn.

Have you always kept notebooks?

Yes, and until I got too busy I used to keep fairly detailed notebooks and then for some unknown reason copied it all out neatly again. So I've got notebooks going back to 1968, a whole set of A4 'neat' desk diaries from then to 1987, then after that I didn't have time to copy it all out.

I've got back into photography since the digital revolution. I took a camera on my world trip, and I look back at the photos and think how terrible they were, but I used to show them to bird clubs, and I honestly believe they thought they were quite good, which I guess they probably were at the time!

Have you ever been into recording bird sounds?

Yes, I dabbled with that quite a bit, made my own fibreglass parabolic reflector because I couldn't afford to buy one, and even won a sound recording competition.

What's your best bird find ever?

The Rock Sparrow I co-found with Richard on 14 June 1981. I remember it as if it were yesterday. Richard saw it first. I was walking along the top of the beach, partly looking out to sea, coming back from Cley's North Hide towards the car park, and he was walking along the inland side of the beach. He stopped because he saw this bird under the fence. He'd still got his giant draw tube brass telescope (about eight feet long!), so he lay down on the ground – eventually he waved to me. He didn't know what it was but it was obviously something good. Initial thoughts were that it was an American Sparrow. Eventually it flitted up onto the fence, fanned its tail as it did so, and then the penny dropped and I realised what it was.

What about your worst dip?

There have been many dips, some of species seen later, others not. At the time, it was perhaps the Pallid Swift at Stodmarsh in May 1978. It was long before pagers – I was up in Norfolk, came back to Nancy's to find a message: 'Pallid Swift at Stodmarsh performing well', went haring down there – and got there just too late. As we drove into Stodmarsh village, everybody was coming away, saying it was still there. We got to the Lampen Wall and there was a big crowd who saw us arriving, and they were genuinely really sorry, looking at their feet, feeling embarrassed, as it had been flying up and down for six hours at head height and disappeared just one minute before we got there. So it was a really painful dip, as it was a first for Britain at the time – but not bad enough to make me want to give up. I saw it a few days later anyway!

Where are you in the top ten now?

I think Mel Billington and I are on about the same number – whatever that is – at the top.

So, you occasionally go on a foreign trip now. Do you feel nervous about missing something here?

I've been lucky I haven't missed much when I've been on foreign trips, but my British list does mean a lot to me.

If it means a lot to you, does that mean you want to be number one, or in the top ten?

I find it difficult to analyse. I know that Steve Webb openly, desperately wants to be number one, whereas it's never bothered me. I just want to see everything, and that's all. If something turns up and nobody I know sees it, it doesn't bother me too much, but if there's what I consider to be a really exciting bird and I could have seen it but didn't, it really annoys me. There is an element of 'all my friends/rivals have seen it and therefore I want to see it', but that's not why I do it – I just want to fill in my British (and Irish!) list. Something like an Eastern Crowned Warbler – that's a bird I've been dreaming of ever since I read about the one on Heligoland in Heinrich Gätke's book donkey's years ago.

You recently decided to close *Birding World*. That must have been a tough decision. Why did you decide to do that? And what were the best and worst aspects of producing it for so long?

Producing *Birding World* was great fun; it was my living too and infinitely preferable to a 'proper job'. I loved being at the heart of the birding news and corresponding with the many excellent birders and photographers across Britain and the Western Palearctic who so generously supported the magazine. But it was very time-consuming too: I worked harder and much longer hours than I ever did in my only 'proper job' (being an optician 1974–81). Like many of our subscribers, I shall miss *Birding World*. I still think its content was great and there is nothing else that fills its place – but not enough to continue doing it. Quite simply, I reached the age of sixty and decided to spend my old age relaxing and birding rather than sitting at a computer screen for far too many hours and stressing about deadlines every month. My partners were of a similar age too, and just meeting that deadline on time every single month for twenty-seven years seemed enough. A few people (including some well-known names) did come forward, with serious interest in buying the business of *Birding World* (despite internet competition, it was still perfectly viable financially), but they wanted to buy the business, not do the work!

Is there somebody in birding history that you would really like to have met?

Heinrich Gätke, the pioneer of migration/rare bird/vagrant studies.

Where is the best place you've ever been birdwatching?

Impossible to say, but Bhutan is up there.

Where is the place you've not wanted to go back to for birdwatching?

Philippines. Great birds but horrendous overpopulation and deforestation thirty years ago; it must be depressingly far worse now.

If you could go birding to one more place in your life that you've never been to, where would that be?

Mongolia.

What is your favourite bird group?

All of them!

What is your most wanted bird?

Macqueen's Bustard in Norfolk would be nice.

You're on a desert island – what piece of music would you choose?

'Dark Side of the Moon' by Pink Floyd.

Favourite film?

Birdman of Alcatraz.

Favourite TV show?

Anything from Sir David Attenborough.

Favourite non-bird book?

Handbook of the Mammals of the World.

Favourite bird book?

The Handbook of British Birds by Harry Witherby and his team.

MARK COCKER

-

Mark Cocker is an author who was born in the 1950s.

INTERVIEWED BY MARK AVERY

Where did you grow up, and how did you get into birds?

Well, I was always into wildlife as a child, and from a very young age my parents used to buy me books on nature, and I think I started to be interested at about six or seven. But I never went birding, so I was a passive, book-based reader of nature 'stuff'. And then a friend from school started birding near my house and I began to go with him at about the age of twelve – I always regret missing out those years between about seven and twelve. Then there was a natural progression from twelve involving the local Field Club, the YOC and then a maths teacher at school who took me, from 1973 onwards, to Spurn to go birding.

School was in Buxton in Derbyshire and we were surrounded by wildlife. As a child, in the area of woodland and moorland that was immediately above the house, I used to see Ring Ouzels, Wheatears, Whinchats, breeding Dunlin, Twite, Short-eared Owls, Wood Warblers, Tree Pipits, Redstarts, Common Sandpipers, Grey Wagtails, Dippers, Curlew and Red Grouse. Lapwings and Grey Partridges were in the fields. Now that list is like a moth-eaten rag. I didn't see those birds every day, all of them, but that was the suite of species I grew up with from about 1972–9. and, you know, many of those have now gone. Cuckoos and Wood Warblers are gone. Common Sandpipers are possibly still there. Dunlin are gone. Redshank are gone. Ring Ouzels are gone! Twite, all gone. It's an astonishing change in my lifetime.

So that upland area was a big influence on me. Spurn Head was the other place we used to go – relentlessly from about 1973–7.

Who was 'we'?

We're all still birding – myself, John Mycock and Mark Beavers. Mark's a former Derbyshire copper and keen local birder and John works in the South Downs National Park. And then I expanded that community to include lots of

Cheshire birders, and Manchester birders and Yorkshire folk. From 1976 we used to go to Cley, and that's really the narrative described in *Birders: Tales of a Tribe*: coming to Norfolk and being obsessed with the Norfolk coast and being obsessed with the culture of twitching in Norfolk. So I went twitching with serious intent until about 1982, and then with moderating or fluctuating enthusiasm until I totally gave up around 2000 or so. And now I couldn't give a damn! No, I mean I really couldn't care less about rare birds. And that's been quite an interesting journey.

In some ways I've come full circle because the Buxton Field Club was a very good natural history group. It introduced me to a very rounded wildlife encounter, and I've returned to that model. Now it's a lot of moths and a lot of flies – and though I'm really bad at it and not an expert in any way, I love looking at all these other things. And all these other life forms are so difficult and absorbing – imagine looking at the hairs on the thorax or abdomen of a fly! – that birds have, to some extent, taken a back seat. Obviously, however, my first love was birding with a pair of binoculars.

What were your first pair of binoculars?

The first pair of binoculars were stolen off my brother. I think they were about £8 from Boots. They were really bad! And the leather strap (which was probably plastic) was stapled together. They had dirt on the lenses – I mean, they were really rubbish! Of course, your eyesight is really good when you are about thirteen and you compensate for this. Now I've got a pair of Swarovskis.

I thought you would!

I've always liked to have decent binoculars. I do love these Swarovskis – I got them off Martin Woodcock. They were second-hand so I didn't pay a fortune for them. But you know – I feel that if you are going to be a birder then you might as well have the best you can possibly afford. People might argue the toss about Swarovski, Leica or Zeiss, but one of those three brands is going to be all you need.

I did have a pair of early Swarovskis – Habicht Dianas – and from the age of thirteen to twenty-four they were the binoculars of my 'apprenticeship'. Very decent bins.

Early bird books?

What was so extraordinary was the paucity of bird books. There really was only *The Observer's Book of Birds* in the early days – that's what I used to go out with, and I probably had two copies over time. And then later I remember getting Heinzel, Fitter and Parslow (*Birds of Britain and Europe with North Africa and the Middle East*) and then there was the Hamlyn guide by Bertel Brunn – the other illustrated birds of Europe. Those were our key guides. In those days, the maps were in the back of the books and I started to dream – on the basis of looking at the maps – about going abroad and seeing amazing things that I couldn't ever dream of clapping eyes on here, like a Red-flanked Bluetail – stuff like that.

How important was school?

We had at least two teachers who took us out birding in a way that would be almost inconceivable today – older men with young boys. The whole obsession with 'stranger danger' has destroyed that apprenticeship process, which was probably fundamental to all young naturalists of our generation. Those young people were empowered to go and see things and do things because of relationships (nothing improper ever happened!) that would be carefully scrutinised today. Does it still happen today? My maths teacher was unmarried at the time and used to take a group of us out to Spurn. My parents were totally happy with this and so was I – and I remain eternally grateful – because it was a fantastic opportunity to get to see things and do things that we'd never have done otherwise. I really lament the current fearfulness and I write about it fairly regularly in *Birdwatch*. But we were the same with our kids – we never let them go out far from home. Looking back on my own parenting, I feel bad. Because I used to go out everywhere – didn't you?

Yes, just the same. The same as you – I went on trips with a couple of masters from school and then in the holidays I'd be out on my bicycle (and there were no mobile phones in those days), cycling to Chew Valley Lake for the day and coming home in time for tea.

You know, we were out in environments which were not mild and gentle. The Derbyshire hills are a pretty rugged place and in some conditions can be dangerous. But we grew up with all that and we felt at complete ease in blizzards, fog, rain, snow. You know, the outdoors is everything to me because of that childhood. And I think we've denied it to our own children and I think that's tragic.

So, you moved on to twitching and north Norfolk, where you and I may well have stood next to each other at some stage.

I'm surprised we didn't meet, though, Mark if you were a regular there. Because I was avid …

Well, you were probably more avid than I was, but when I was still at school, I used to be in north Norfolk for the last week in August, hoping for a fall of Wrynecks, Barred Warblers, Icterines ...

... which of course there were at that time. Then, we didn't think very much of seeing Barred and Icterine Warblers, now I'm not so sure that they occur ...

Not so much ...

I can remember logging 300 Pied Flycatchers in Wells Wood in Norfolk and not thinking a great deal about those numbers, but obviously really relishing it. I can remember one walk we had to Blakeney Point and I'm pretty sure we had ten Wrynecks. And we saw the fag end of that suite of annual autumn falls which were well recorded from the 1960s which started in Suffolk – was it in 1967? – with astonishing numbers of migrants. We used to have numbers that made us feel that we were witness to that same process. I am not sure it even happens today. That's the sort of thing that gives you pause for thought. I cannot remember when I last saw a migrant Pied Fly.

You wrote about twitching and the people involved in it in *Birders*.

When I wrote *Birders* I'd already given up twitching – it was published in 2001 and written in 1999–2000. So it was in some ways a retrospective, if you like, a 'looking back' on a period of my own life that I relished as a memory but was losing as a living activity. I felt a little uneasy that I was narrating a story that I had ceased to be a part of. Since then I've moved on a long way, and *Crow Country* is a kind of riposte to *Birders*.

Birders was received very much as a book about twitching, but it wasn't really about twitching. Primarily it was about people who enjoyed (or suffered!) a total absorption in birds – people who were completely fixated with feathers. What I tried to argue in the book was that twitching formed part of a spectrum of activities that a birder could pursue, so people like you, Mark, could be keen on rare birds but also an avid local patch worker, trying to find as many birds as I can on my local patch. I tried to present birders as people whose lives were shaped by birds. I wanted to describe how the human heart could be moulded by birds. That was what the book was really about. Alas, now I have a reformed smoker's attitude to twitching and, if I'm honest, it has become banal and meaningless for me.

Harmless?

Well – largely harmless. It seems to me that there is a distinction between being a 'lover' of birds and being obsessive about birds. If you love birds in the same way that you might love a person, then the object of your love comes first. But with some birders I wonder if the thing that comes first for them is their own obsession. I question whether some keen twitchers genuinely feel much for birds, because if they worked through it all they would become deeply troubled by expending all that carbon and they would become more attuned to the idea that *all* birds are equally meaningful. After all, wonderful birds are all around us. It's not just the occasional oddities that are beautiful or fascinating. Does one really need that rarity element to make the experience meaningful?

I am as moved by the sight of any common bird as I am by a rare bird. There's no difference between the affecting power of a Eurasian Wigeon and an American Wigeon, or between a Common Teal and a Blue-winged Teal. I am as fulfilled by a Blackbird as I am by an Eye-browed Thrush or Naumann's Thrush – you know, those things that I've chased in the past. I feel I have purged myself of that. Unfortunately, there's a sort of moralistic, preachy quality to my views now!

In *Birders* I divided birders into categories from Robin-stroker to twitcher – I have now definitely morphed into a Robin-stroker! For example, I fed my birds in the garden this morning! Of course, I'm not *just* a Robin-stroker – I do realise that the Blue Tit in my garden isn't really 'my' Blue Tit and that it isn't the same one that has been there for years, but Robin-strokers are where it's at! Robin-strokers are the only hope for nature because there are so many more of them. They are the environmental community!

From twitcher to Robin-stroker, that's the journey I've made. I think we're all seeking for ways of being members of a modern, affluent society but adjusting that to our passion for nature – and that's what my life has been for about fourteen years.

You spend most of your time writing about nature. Why is that?

I was not very employable. I talk a lot and I don't think I'd have fitted into an organisation very well; I didn't take to restraint and control. But I did work for the RSPB briefly and for English Nature and for BirdLife International. I was a warden for the RSPB – I wardened the Parrot Crossbills when they bred at Holkham in 1985 and then I got a job with English Nature as a summer warden for two years, also at Holkham. At BirdLife International I did some freelance work as a precursor to the Important Bird Areas project.

But I really wanted to be a writer: my two main interests were wildlife and English. I studied English at the University of East Anglia. It felt as though I worked in two compartmentalised worlds – the world of literature, poetry and philosophy and the world of birds – and I suppose that eventually I found a way to combine those two. That was my unconscious quest: for a life that combined my various interests – coupled with a general unsuitability for employment!

I didn't write about nature when I started. I wrote biographies of people who were naturalists but who also had wide hinterlands of intellectual activity – so Hodgson (*A Himalayan Ornithologist: The Life and Work of Brian Houghton Hodgson*, with Carol Inskipp) was a scholar of Himalayan Buddhism and culture, while Richard Meinertzhagen (*Richard Meinertzhagen: Soldier, Scientist, Spy*) was a military figure and a spy as well as a liar of the first order. My subsequent books – a study of travel writing and travel writers (*Loneliness and Time: British Travel Writing in the Twentieth Century*) and a history of the destruction of tribal peoples (*Rivers of Blood, Rivers of Gold: Europe's Conflict with Tribal Peoples*) expressed my wish to be a 'real' writer, but I was drawn back to wildlife and for years I've done nothing else except write about my fixation with nature, the role of nature and the importance of nature. It's been an unconscious journey to bring all of my interests into one place, I suppose.

What is the life of a writer like?

There is flexibility in being a writer. That's one of the things that I sought. But you have to be adaptable; you have to do many things. I had to do a lot of things that I don't think I was that good at – like tour leading – to make a living. I wasn't a bad tour leader – but I just wasn't a *great* tour leader, whereas I knew other people who had the temperament for that type of work. I later became concerned about my carbon footprint and those sorts of issues. But my life has been very varied.

The other determinant of a freelance life, which you will have encountered, Mark, is that money doesn't come easily – at least, it didn't to me. I'm hopeless at making money! So there have been moments of sunlit relaxation, but set amongst a relentless concern for where the next pay cheque is coming from. And squabbling, and moaning and bitching about how little one can get paid for what you do!

I've also been able to do some really lovely things, don't get me wrong. I think I've written at one time for all the major British newspapers, but especially the *Guardian* and these things can be more lucrative and they subsidise other parts of your writing career that are not so well paid. The aim is to have a broad, varied portfolio of activities in order to reach different audiences.

The pluses are control and flexibility. If you want to go and see a Humpback Whale on the east coast, if one turns up, or a rare bird, in my former incarnation, then one can. But you also have to be very disciplined – there is no real scope for mucking about and staying in bed; you have to get up early and get on with it.

I'm not a natural writer so I do a lot of, let's call it, frigging around, to get the words out. But I don't write much more than a thousand words a day.

That's quite a lot!

Well, it's enough. Usually, when I was writing *Birds and People*, I wrote about 10,000 words a month (2,500 words a week) of finished text. It was a slow process.

I think there is an element of glamour that attaches to being a writer – you are asked to speak, your opinion is sought, people ring you up – but the reality isn't very glamorous. It's often rather mundane. But it's very pleasurable. And it introduces you to interesting people. I know lots of writers, and generally they are interesting people. Some are psychopathic, but a lot of them are wonderful, creative people. It's great to be with creative people. And that's a privilege. That's one of the joys of being a writer.

You must be enormously satisfied with *Birds and People*?

When you are an author you are looking for challenges. I knew that *Birds and People* was the biggest mountain I would climb in my career.

How many words was it?

I think it is 430,000 words.

How long did it take?

We started work on it in 2005 and it was published in 2013. We signed contracts in 2006. I started researching in 2007 and I spent a year simply reading around the subject. I started writing in 2008 and finished in October 2012. It was very slow progress, although I kept all my columns going as well. *Birds and People*'s select bibliography is 750 titles but that only includes the titles that are quoted in the footnotes. If it didn't end up in the footnotes it isn't in the bibliography, and so I sieved at least another 500 books that don't get a mention!

It was a very, very large task – and it was almost too much for me. But, you know, the scale of the challenge is part of the sense of achievement, although I don't spend much time dwelling on it. You move on to the next thing.

It's been a critical success, but it hasn't been a commercial success, at least not as much as it could have been. It's sold around 20,000 copies so far, I think. But it is certainly a marker – it's a marker of what I can do, with David Tipling, of course, who was one of those creative people who was a delight to work with. It was, to some extent, a statement of what I could do.

If someone asked you to do it, what about volume two of *Birds and People*?

I'm at a point now that, if an American publisher came to me and said, 'Here's a million dollars, and we will fund whatever you want to do if you'll write *Birds Americana*', then I'd think about it. It would be a similar scale of project. Have I got the stamina for that, involving moving to the States?

The look on your face says you have!

Well, I certainly didn't last year, or most of this year, but now, yes, I'd probably think about taking it on.

It would be great fun, wouldn't it?

It would be a great book to do. I don't think David and I will ever be asked – to get Brits to do *Birds Americana* might be a bit more than the USA could take.

I'm looking to a different kind of writing now, the next two books are short.

Birds and People has been a critical success with almost uniformly good reviews. We were listed in, I think, thirteen round-ups of Books of the Year, including those by Andrew Motion and Jim Crace, and we have been asked to speak all over the world, which is good fun. It's one thing that would always go on my biography because it's so big, and I like to think it is an important work – it's an affirmation of the importance of Robin-strokers! Birds mean so many things to so many people, and one of the things I lament is that some mainstream birders, even those working in conservation, have a problem in understanding how diverse and important are these other ways of engaging with birds. To a large extent, keen twitchers would not be contributors to *Birds and People* – often it was what you might call 'ordinary people', such as farmwives at their kitchen sink, who said the most memorable things; things that move you deeply; things that no twitcher would think of saying. And I think that's a power of the book: it gave voice to all sorts of different people. A hugely diverse cross-section of humanity got to speak through its pages. It's a global chorus of why we should care about birds.

What's next? How can you follow *Birds and People*?

You park it and forget about it. Your next blank page has to be filled with material that is worth reading. I've got a new book coming out which is called *Claxton – Field Notes from a Small Planet*, which is a digest of pieces I've written in *Birdwatch*, the *Guardian*, the *Eastern Daily Press* and so on, and it is about my relationship with place – this place where we are now. It's the part of the relationship with birds that isn't manifest in twitching. The idea is that a bird – all wildlife, in fact – becomes more meaningful when encounters occur in one particular place. Claxton's been my home for fourteen years and the book is a statement of the importance of place. So that comes out in October 2014.

I'll look forward to it. Do you have a favourite film?

Lonesome Dove.

A favourite non-bird book?

How Tom Beat Captain Njork and his Hired Sportsmen by Russell Hoben.

Favourite TV programme?

Big Bang Theory.

Favourite music?

'It Never Entered My Mind' by the Miles Davis Quintet.

IAN WALLACE

Ian Wallace is a writer and artist. He was born in the 1930s.

INTERVIEWED BY KEITH BETTON

When did you first become aware of birds?

I was four. In a Shetland summer, before the Second World War. My dad showed me a Puffin.

We had a launch called *The Skylark* and he took me on trips to wonderful places like Noss and Bressay. Among my strongest memories was a Killer Whale going right underneath our boat! I also remember looking out of a window of the Queen's Hotel in Lerwick and seeing Arctic Terns hovering just outside.

You often wear a kilt and a Balmoral bonnet, so are you truly Scottish?

Oh yes – 100%. My dad was from Ayrshire, and Mum from Sutherland stock, and although I was actually born in Norfolk, I went to boarding school near Edinburgh.

Who influenced your early interest in birds?

The first person would have been John Pirie, who walked along the Shetland beaches with me and drew very helpful pictures of the birds in the sand using a stick. Then at my prep school – Dalhousie Castle (now a hotel) – my French master, Mr McGregor, was interested in birds, and more importantly the groundsman told what he'd seen. I well remember that one day he showed me a dead Woodcock. He removed a pin feather from the bird's carpal joint and I used it for fine line painting for years.

Did you keep a diary?

Yes, I kept lots of notes from 1943 (when I was ten), but sadly I have lost the early ones. In 1947 I went to Loretto School near Musselburgh. Once again it was a French master, Mr Turner, who encouraged me to put my observations into the log of the Loretto School Ornithological Society. Birdwatching was really popular: 50 out of 200 boys were members! We used to have film shows and everyone wanted to see them. I clearly developed my literary bent then, as I edited the school magazine and the Society's bird report! My own diaries from 1951 now take up four feet of bookshelf at home.

What was your first big bird discovery?

I and two other lads were very lucky on 13 May 1950; we were at Aberlady Bay and chanced upon a Lesser Yellowlegs. Naturally, we didn't know what it was: we were all very familiar with Greenshanks and Redshanks, but this bird had a square rump patch which I noted very carefully at the time. This was only the second Scottish record, and it was accepted by Bernard Tucker, the editor of *British Birds*, then the sole judge of such claims.

Did your family encourage you in your hobby?

Certainly – Dad was a hill-walker, and although he wasn't really a birdwatcher, I badgered him to get me to Fair Isle for a week in September. So, in 1951 he took me, and it was a life-changing experience. Apart from seeing amazing birds like Black-headed Bunting, I was meeting great tutors – such as Ken Williamson, who was the observatory's first director, and Maury Meiklejohn. The end-of-day log session was revelatory of bird recording disciplines.

What happened after you finished your schooling?

In December 1951 I became subject to National Service and I was trained by the Royal Scots. At that time many soldiers were being despatched to Korea, but instead I was sent to Kenya. I served in the King's African Rifles from 1952–4, based around Mount Kenya and near Lake Nakuru, both wonderful for birding. Apart from getting to know African species, I enjoyed spending time with Palearctic migrants in their wintering grounds. I kept plenty of records, and this led to my first foray into publishing joined-up observations: a paper in *British Birds* on the racial variation of Yellow Wagtails in Kenyan habitats. It appeared in 1955, five years after my first note!

Were you as keen on birdwatching when you returned to the UK?

Most definitely, and in 1954 I was back at Fair Isle, and my interest was strengthened on meeting great birdwatchers like Horace Alexander, Guy Mountfort and James Ferguson-Lees.

What happened next?

In October 1954 I enrolled at Clare College, Cambridge to study economics and law. At that time the university was full of keen birdwatchers such as Chris Smout, David Ballance, Clive Minton, Bill Bourne and Ian Nisbet. We were all involved in the Cambridge Bird Club, and for my last year I was President. Then in 1957 I started work in London. The London Natural History Society was very strong, and the likes of Stanley Cramp, Phil Hollom and Dick Homes made sure that I was kept busy with the London Bird Report. In 1960 and 1961, Tony Gibbs and I organised trans-London migration watches and estimated the first autumn passage at four million birds.

Did you carry on visiting other places for birdwatching?

Yes – and in particular St Agnes in the Scilly Isles. Back then people hadn't really discovered the significance of the whole archipelago for finding rare migrants. So John Parslow, Keith Hyatt, Brian Milne and I developed the St Agnes Bird Observatory. As it was near to the Bishop Rock, we assumed that all the rare American vagrants would land on our chosen isle. Little did we know then that they landed on all the others too!

In the 1960s you took part in the famous expeditions to Jordan. What was that experience like?

Amazing! We went there three times – in 1963, 1965 and 1966. We worked really hard to record everything we saw; it often took us three hours in the evening to carefully transcribe each day's observations. The expeditions were hugely significant for Jordan's ornithology and led to an attempted system of national parks.

What was your local patch at this time?

It was Regent's Park in Central London. I first got to know the area in the late 1940s, but from 1959 to 1965 I visited the park most days. It was effectively my back garden. I was particularly interested in the visible migration over Primrose Hill, and there were often impressive falls of migrants in the park. One morning, I counted thirty Turtle Doves that had dropped in briefly!

Why did your studies there end?

Well, I was posted first to Gloucestershire – in the winters of 6,000 White-fronted Geese – and then overseas as marketing director of Nigerian Breweries. So from 1968 to 1971 I was based in Lagos, and my work regularly took me to Ghana and Sierra Leone. My best adventure was travelling by car from Lagos to Lake Chad and back – a total of 3,333 miles in 12 days. I had several crates of beer in the boot so whenever I was stopped by a troublesome official it was easy to provide them with a small gift to make sure I wasn't delayed too long from the next bird!

You got involved in the British Birds Rarities Committee at this time. What was that like?

I first joined the team in the 1960s and then again when I returned from Nigeria. I was Chairman from 1972 to 1976. I think it is difficult to realise just how little we knew about identification forty years ago compared to today. I enjoyed being part of the panel, but I didn't see eye to eye with some members. My style of writing descriptions was somewhat more artistic and jizz-ful than most, and when I worked out that 13% of my own claims were actually being rejected, I realised that it was time to move on! Arguing with zealots is a waste of time.

How did you develop your own artistic style?

I have always kept notebooks and I love doing sketches. I struggled with watercolours, so now I use gouache. My illustrations are always of birds in action – flying, feeding and looking alert. If I can, I try to use my art to tell a story. I just can't draw birds sitting still and doing nothing. Bob Scott once noted that artists see things that most birdwatchers simply miss.

How come you moved to Yorkshire?

In 1972 I was offered a job in the fish industry in Hull and so I followed in my father's footsteps. I spent many happy days at Flamborough Head, getting it back on the ornithological map.

Your first book was *Discover Birds*. What made you decide to write it?

That was in 1979. I was bored with the rat race, and I wrote my creed for joyous birdwatching in just three weeks! Many people seemed to buy it but, despite that, the publisher went bust! Hence I went back to a proper job in charity fundraising. I have written several other books, and in 2004 I decided to share my thoughts on birds, birdwatchers and birdwatching in *Beguiled By Birds*. A revised edition will appear in 2015. I also enjoyed doing my monthly column for *Birdwatching* back in the 1980s. It was amazing just how many people read it.

You moved to Staffordshire in the 1980s. That must have been a very different birdwatching prospect, surely?

Yes – completely different. I love the coast, but I also love patch work, so I decided to concentrate on the latter. From 1985 to the present day, I have kept detailed notes on the farmland birds that surround my home near Burton-on-Trent. I particularly enjoy talking to the gamekeepers and shooters – they know so much. You may think that inland counties don't offer much variety, but visible migration keeps me fascinated and I keep finding good birds too. I am sure I have seen several Oriental Skylarks in amongst the autumn flocks of Skylarks. Sadly, two of my claims have not been accepted by the Rarities Committee, but believe me, someone will find one soon!

Talking about the future of birdwatching compared with your experiences of the past, what are your current thoughts?

I think the kind of legacy that I inherited from excellent teachers – people who took us into the field, showed us the wonder of birds – and which has always kept me going may no longer exist. I had a most interesting conversation with Andy Gosler recently. He works in ethno-ornithology, and he told me they had done some research in Oxford with biological students – so not little kids. Questioned on five common species of birds, mammals, trees or whatever, their level of recognition was minimal, which is incredible. So the people coming to do a biology degree have not had the grounding from a similar chain of mentors to those who took me forward.

It's very difficult to feel too pessimistic when you stand in the middle of the BTO conference, but when I look for the old connections away from this sort of caucus I don't see them. I worry too about the whole 'VIP celebrity/ soundbite' connection that has been bolted on to the hobby. I don't think that generates the more than seventy years of private joy that I've had. People have got to understand that our connection with nature is at risk.

I'm particularly critical of the major NGOs for not doing joint socio-demographic and attitudinal research into this problem. I don't even have confidence that the RSPB really understands how many million people would turn on if we could ring the right bell for them. I don't sense anybody trying to market what nature pays you back, mostly for free, once you get out into it. I don't think armchair TV does it. Where is the invitation to get into wellies and out into the nearest muddy field or decent wood? That's what you have to do to get face-to-face with other beings and then get your own personal joy from them.

You talked about the RSPB there, and mentioned nature. The RSPB has just changed its magazine name from *Birds* to *Nature's Home*, so it has obviously identified that its future may be better if the RSPB is aligned with nature as a whole rather than simply birds.

I think it's an extremely dangerous change of tack, because at a time when there are other, even new, environmental charities doing well with membership growth, the RSPB has been stalled for nearly a decade. I don't know that I'd go for a broader appeal in that situation. I think I'd try to stay sharp and get back to where the RSPB started being really passionate about the collapses of so many birds as the key indicators of the lost commonwealth of nature and man.

You said you'd seen 398 accepted species so far in Britain. There will be people today who've seen that many in three or four years of starting birding.

That's what Lee Evans says: they get to 400 in three years, then burn out as lister/twitchers and go and do something else, apparently! Which makes it sport – and I fully understand sport, I've followed Manchester United for most of my life. But do you want that, or real payback, from your citizen science and a feeling of contributing and doing something to stop the rot of the poor planet? If you do want the latter, and I think kids can still be inspired by the cause and stick at it for a lifetime, then simply wanting to get your first 400 is not enough. Indeed, if you split everything to the maximum, you can get to 654, sheer nonsense!

We've seen some good presentations at recent BTO Conferences, and some of the presenters have been in their twenties, but I don't see many people under fifty in the audience. What do you think about that, if this is the future of birding we're looking at?

I'm slightly encouraged. I do think the audience's average age is a bit down, and the young professionals are a delight to be with. Better still, some of them are ladies, so we've got a less masculine obsession in the way we deal with enumeration. I think the work Chris Wernham did with *The Migration Atlas*, and Dawn Balmer with the latest *Atlas*, is just tremendous. What's so interesting is that zoology students of thirty or forty years ago had to go and work in the civil service, as there were no zoology jobs. It's a delight now that there are sufficient funds for at least some of them to have their lifetime skill turned into a lifetime career, and they're doing bloody good work.

There has been quite a change in field guides in the past five years and several have more use of photos, but there are field guides which have montages, following the Richard Crossley approach – have you seen those?

I reviewed the first, and I've seen the others. If I were still collecting bird books for identification, I'd probably buy them because as a photo database they're excellent. I don't think they tell you much about bird identification in terms of learning it as a mental process and a field skill, though.

I thought you might be a bit more positive, as in your books and paintings you quite often show birds doing something you might not expect, such as flying away, and Richard Crossley will show birds flying away rather than towards the camera, as they might do in real life.

I'm not unattracted to the concept. What I said about his *Eastern Birds* book was that he packed so much in, and there were so few connections with the text, that for beginners it was quite a leap to match the myriad images to the clues in the text. At least dear old Peterson, Mountfort and Hollom – which I had quite a lot to do with – put diagnostic marks in italics – you couldn't miss them!

Do you have any worries for the future of ornithology and birdwatching?

Only the inheritance of citizen science by sufficient young people, I think that's the big problem; the BTO universe is also going to die out in ten or twenty years – who's going to replace its loyalists slogging around the 2,500 Breeding Bird Survey (BBS) squares or whatever? I won't be alive when the next *Atlas* comes out, but that will tell us about future allegiance. The recent *Atlas* score of 40,000 contributors, particularly the core 18,500 almost all directly inputting online, was amazing. I would not have believed that would happen. They didn't keep precise figures of the human effort behind earlier *Atlases* (10,000 to 15,000 was about the best, and that was the first one), so this 2007–11 *Atlas* was a cracking effort.

Maybe I am too pessimistic, but I am also saddened by the loss of *Birding World*. It leaves a gap in our avian literature. Will others fill that gap, or just tweet and twitter on?!

I expect your first telescope was a Broadhurst–Clarkson or something like that?

It was a highland stalker's small telescope, a make I don't remember. I think it got me my first Sanderling. The interesting thing is that when I finally worked up to Swarovski bins, they were stolen from my car, and I haven't missed them because the birdwatching I do is all patch work and I rarely raise my glasses, as I know the bird either by eye or ear. My first thought was, 'Am I insured, and where do I get the next pair?', but to my Scottish delight it just ebbed away and I'm not going to spend the money!

You are wearing an American device round your neck. A bird calls, and you can record the call. The device will collect the sonogram, analyse it and maybe give you five options of what that bird might be. Do you see a time when binoculars will have an image reader: you see a bird and you'll get some options for what the image might be?

I don't see that because I'm such a personal perception addict. But James Ferguson-Lees forecast just such things, and if they facilitate bird identification, then you can't have an old crud like me saying it's wrong!

Is there anything else you'd like to say about birding today that you think might be of interest?

The two people I like best in my patch are both gamekeepers. They don't shoot vermin, they control it humanely, and both of them give hundreds of pounds' worth a year of peanuts and nyger seed to wild birds in the middle of their pheasant shoot. I just adore them because they are doing more than the undisciplined subsidies that we are wasting our tax on. So you want to ring up somebody who knows? The Game and Wildlife Conservation Trust – it is more in touch with reality on the ground than many RSPB employees appear to be.

If you could meet a birdwatcher from years gone by, who would you choose to meet?

I would like to have spent some time with George Stout. His nickname was 'Fieldy', and he was Eagle Clarke's assistant on Fair Isle back in the early days well before I first visited. There is no doubt that he was an amazing natural birdwatcher, using his practised eyes to spot differences without any access to the field guides that we all take for granted today. It would have been great to have compared our perceptions!

Where is the best place you've ever been birdwatching?

St Agnes in the Isles of Scilly.

Where is the place you've not wanted to go back to for birdwatching?

Azraq in Jordan. It used to be great, but now it's ruined.

If you could go birding to one more place in your life, that you've never been to, where would it be?

Somewhere not too far away, where I could see White-faced Storm Petrels! So probably in a boat off the Cape Verde Islands.

What is your favourite bird group?

Redpolls – I love them!

What is your most wanted bird?

White-faced Storm Petrel.

You're on a desert island. What piece of music would you choose to take with you?

'Johnny Cope' – played on the bagpipes with kettle drums.

Favourite film?

Kidnapped.

Favourite TV show?

Civilisation – the thirteen-part series from 1969 by Sir Kenneth Clark.

Favourite non-bird book?

Birds of Siberia by Henry Seebohm. I know it's a bird book, but I read it for all its fascinating facts on Russia.

Favourite bird book?

Natural History in the Highlands and Islands by Frank Fraser Darling.

ANDY CLEMENTS

Dr Andy Clements is the director of the British Trust for Ornithology (BTO).
He was born in the 1950s.

INTERVIEWED BY KEITH BETTON

What do you remember about your very first birdwatching experience, and how did you get into birdwatching?

I must have been under the age of nine because the first experience I can remember was when we lived in suburban North London in Sudbury. My dad worked on the railways so it was kind of a railway house in quite a nice area, but I can remember sitting at my bedroom window and hearing and watching Swifts screaming round the garden. That's the first time I noticed birds, I think.

And you didn't have binoculars?

I didn't, no. My first ones were given to me on my tenth birthday and they were terrible. They were Dixons Prinz Luxe or something: great big heavy 10×50s, but I didn't know at the time they weren't very good – I thought they were fantastic.

How did you know the birds were Swifts? Did anybody tell you?

I've no idea how I knew. Somebody must have told me. There were two old ladies who lived next door to us and it turned out later that they fed birds in the garden, and it may well be that they said they were Swifts, or I might have looked them up.

We didn't have bird books. My parents weren't interested in natural history at all, but they were intent on education, as a lot of that generation were, and we had the *Encyclopaedia Britannica* and I know I looked up Swifts in that.

No one in the family was a birdwatcher, and there was no one to influence you?

No. My dad moved with the railway to Derby. We went from north London to a Derbyshire rural village and I had two friends whose parents ran a smallholding and a nursery. I remember spending all my days out with those friends in the countryside. You wouldn't describe them as birders, but we found our first Coot nest together just because we were out in the countryside, so that germ of interest was slightly continued through that period.

Were you tempted to take the eggs?

Oh yes, we did take the eggs from the Coot nest, and I can remember knowing it wasn't really the right thing to do, but justifying it after the fact by saying that we only took one or two of the four or five eggs that were there.

Just for the record, this was after 1954?

Yes, it was, but it was no different from what lots of other boys were doing at the time.

When were you born?

1954.

What about school?

Before we left north London I took the entrance exam and got a scholarship to the Haberdashers' Aske's Boys' School, but as we were moving away the scholarship wouldn't have counted. My parents persuaded me I should still go to that school, so I went as a boarder from Derbyshire to London.

There was a Bird Club there, wasn't there?

I formed the Bird Club there under the mentorship of Barry Goater, so he is the person in my life who nurtured my interest in natural history. It rocketed as soon as I got to Haberdashers' and came under his wing, and my very best birding friend is Rick Goater, his son, who I have known since those very first secondary school days. We did also look at general natural history, particularly moths and plants, with Barry.

So you got binoculars for your tenth birthday?

I also got my first bird book, the *Collins Pocket Guide to British Birds* by Richard Fitter and Richard Richardson. It had a set of colour plates of perched birds and black-and-white plates of birds in flight and plumages, and that book and the binoculars were my weapons of use at the time.

Mum and Dad liked to go out for a drive in an evening or at weekends, and they took me to Blithfield Reservoir in Staffordshire, about half an hour from where we lived, where I saw Great Crested Grebes which I identified from the Fitter and Richardson book which said that there were fewer than 1,000 pairs in Britain so I thought they were very rare, and I remember sitting in class at primary school, where you say something you were doing at the weekend, and I said I went to this place and saw a very rare bird. The teacher asked me what it was and I wouldn't say, as it was a secret and had to be protected! So that was a significant event for me.

Were you the only one in your class interested in birds?

At primary school I was the only one, and I can remember painting a Gannet, copying it from a picture and putting it in a school art show. But when I went to Haberdashers', straight away there were one or two people also interested in birds. Rick was at school nearby and I met him at Hilfield Park Reservoir, our local patch near the school. Together we identified two White-winged Black Terns. I did a lot of birding with him and his dad, and we did trips with the Nature Club.

Is it true to say you saw White-winged Black Terns before you saw Black Terns?

I did. And I think a lot of birdwatchers have that journey where they don't necessarily see everything in the right order. And I think – this is me being opinionated – this can be extreme for today's twitcher. People will know I'm a sometime twitcher, there's no point pretending I'm not, but for me there's an important grounding in patch birding and just enjoying local birds, whatever order they come in.

What's your British list?

464.

When did the Bird Club start at Haberdashers'?

Probably during my first couple of years there. We used to be easily able to fill a minibus when we went on trips, but it kind of grew slowly. Those memories are probably later in my school life, though, and maybe four or five of us would go off in Barry's and another master's car to see things. I went to Haberdashers' in 1966 but it was April 1971 (so the trips didn't start in my first year – it was probably the third or fourth) when another biology teacher, Derek Swann, his mother lived at Blaxhall in Suffolk, took me and Ian Saville, a friend of mine, to Suffolk for the weekend. Ian and I stayed in the youth hostel at Blaxhall and Derek stayed with his mother. He took us to Minsmere. That was my first trip to a major nature reserve, and it blew me away. We saw three Marsh Harriers that day, two males and a female, and at the time they were the only ones in Britain.

Did you meet Bert Axell and Peter Makepeace?

Yes, they were there at the time. And we saw lots of common things, for example Gadwall, Shoveler and Avocet, that I hadn't seen before. That was just the beginning of visiting significant places where we could see lots of birds all at the same time.

Were you still using the same pair of binoculars at this stage, and had you got a second bird book by now?

Yes, I had the Peterson, Mountfort and Hollom I remember it had a red band along the bottom. I had that at about the age of thirteen, I think.

In 1971, my parents moved again with my father's job, this time to Scotland. He worked in Glasgow and we lived on the Ayrshire coast at Alloway just south of Ayr. I was still at boarding school and went home for the holidays and went birding on my own at Doonfoot, finding things like Greenshank, Snow Bunting and my first Velvet Scoter. It was nice birding as I could walk there from home – a real local patch.

Did you join the YOC?

Yes – I can't remember the exact year but I was probably about ten and I remember getting the Kestrel badge and waiting for it to come through the post.

Did you ever write notes for its magazine *Bird Life*?

I don't recall doing any. I wasn't a confident boy, so wouldn't have thought I could do that.

Any other good sightings around Haberdashers' apart from the White-winged Black Tern?

In 1972 we found a Hoopoe. The Bird Club had two or three meetings a year at Hilfield Park Reservoir. This was 5 May 1972. It poured with rain all morning and we walked all round the reservoir. Barry was leading, and there were about ten boys. I don't think Rick was there. We'd hardly seen anything and right at the end I walked up to this bush, a Hoopoe flew out and landed on the grass. I'd never seen one before, and it was very exciting.

When was your first overseas birdwatching holiday?

At the end of my second year at Bangor University. I did a huge amount of local birding there. I ran a competition with another guy in second year to see who could go out birding the most days. I re-energised the Bird Club there and we had a foolscap notebook in the library where everybody recorded their sightings. I didn't have a car in those days; I used to hitch everywhere in north Wales and on Anglesey. Second year exams finished by the beginning of June and I went off to the Camargue and the Pyrenees for two weeks. That was my first birding experience outside Britain.

I was one of those birdwatchers who was a bit nerdy about birdwatching in Britain and not wanting to go abroad. We never had family holidays abroad, so I was kind of 'stuck' in Britain.

Between the ages of fourteen and eighteen you said you weren't terribly confident. Were there other boys at the school who told you were a cissy for being interested in birds – and how did that affect you?

Yes, they did. It's part of the whole bullying thing and it's not great, but going out birding meant I could escape from it anyway. At the very early stage I wasn't a sportsman: I was small, I had National Health glasses, I had a Derbyshire accent and I was bullied and teased a lot about that – more about that than the birding, in fact. But then I broke all the school records for swimming and becoming a sportsman stopped all that.

A year after going to Minsmere my parents said 'Bring Ian Saville [my best friend from school] home for the holidays, and we'll go up to the Highlands.' They rented a cottage by the old railway station at Nethy Bridge. We stayed there for a week, and it was like heaven on earth – there were Capercaillie and Crested Tits, we saw my first Long-eared Owls, there were Hen Harriers, and the Ospreys at Loch Garten. It was a really wonderful trip and I thought birding in Scotland was very exciting.

Did you meet any birders in Scotland, such as George Waterston or Ken Williamson?

No, I can't remember who was the Loch Garten warden then. But we had a wonderful time and saw loads of stuff. My parents only stayed in Scotland a very short time and then they moved back to Derbyshire. I did very well in biology A-Level and I wanted to go to University College of North Wales in Bangor to do zoology, for which I was accepted. In 1973, before I went to university, a few of us went to Fair Isle for the first time. We went for two weeks in September and camped in the Pund, which was a croft in the middle of the island, the only place you could camp. A lot of the birdwatching names of the future were there: Mark Brazil, Steve Whitehouse and Phil Gregory. My parents very responsibly said I had to come back a week before I went to university, and to my eternal regret that's what I did. Although we had a great couple of weeks and saw a lot of birds which were mostly scarce migrants – Barred Warbler, Pectoral Sandpiper, Bluethroat, Rustic Bunting – the day after we left there was a White's Thrush and a Lanceolated Warbler, neither of which I've subsequently seen in Britain. So the Fair Isle trip was fairly significant. I really loved it up there and I wanted to go back.

A significant birding event from school days was that, in 1972, towards the end of my school career, we went to Shetland with the school Bird Club, and saw Red-necked Phalaropes both on mainland Shetland and on Fetlar. But, more significantly, we found a Snowy Owl on Ronas Hill on mainland Shetland before we'd seen them on Fetlar where they were breeding. So that was a fantastic trip. I can remember Barry running a moth trap at Baltasound right in the north of Unst, and a local man driving past in the middle of the night, seeing the light and driving his car right off the road into a ditch – we all had to help him get out. So those types of trips were very formative in making birdwatching an OK thing to do.

What size of British list did you have at that time?

Getting on for 300, maybe.

And were you a twitcher?

I'd done one or two twitches. The first I remember was in July 1974: a Ross's Gull at Stanpit Marsh in Dorset (then in Hampshire). It was a juvenile, it had a 'w' on its wings and a neck ring. I saw it because Phil Vines, who was a London birder, was friends with Ian Saville, my best friend. We were all going for a weekend's birdwatching to the New Forest to look for Woodlark – the night Phil turned up he said, 'We're going now and we're going to be at Stanpit at dawn for the Ross's Gull.'

So then I went to university in Bangor. My parents asked me what I wanted to do for my birthday at the end of October. I'd only been at university three weeks, but I went home and said I wanted to see the Buff-breasted Sandpiper at Shotton, which we did – in other words, halfway back to Bangor – but I guess I was beginning to get bitten by the bug of seeing rare birds and needing to travel to see them. When I was at university doing zoology, it was a fantastic life. I was quite 'outdoorsy': I liked mountain walking, I was a climber, birding was a big deal, and north Wales was a great place to be. I didn't have a car, but I could hitch all over Anglesey finding my own birds, including a Long-billed Dowitcher there, and a Short-toed Lark on the north Wales coast. I did some Peregrine wardening at Llanfairfechan and we watched the eyrie. One day we were in a lay-by and heard this bird singing – no one knew what it was but it turned out to be an Icterine Warbler.

In 1974 and 1975 I went to Fair Isle in the autumn again. 1974 was fairly uneventful; I think I went on my own and saw the Black-browed Albatross at Hermaness on the way. In 1975 I went with Ian Saville and Phil Vines – it was a dramatic year because there were two Tennessee Warblers on Fair Isle and we saw them both. Tim Cleeves was there and that's when he met his future wife Ann, who was the observatory cook, and Pete Roberts was also up there; he was the assistant warden.

What were you planning to do when you left university?

At that stage I didn't know, and a PhD landed in my lap because I got a really good degree. I did my PhD on Animal Behaviour of American Mink, studying their ability to hunt under water without physiological adaptations to the eye.

I designed an animal behaviour experiment where we could see how well they saw movement in air and under water. We filmed them catching fish in a pool and discovered that the fish had to move before the Mink could see them, whereas if the fish stayed still while an Otter was hunting it, it became dinner. During that time I moved to Durham because my supervisor moved there, and I had no reason to stay in Bangor.

I did a lot of twitching while I was supposed to be writing up my PhD. I fell in with the Teesside guys, particularly Tom Francis and Martin Blick, and Tom remains a very good friend to this day. I still see him on Scilly every year and he's a great guy. He worked at ICI, he was a nice family man, and a really good birder, really well respected and well-travelled. Between 1979 and 1981 I was really into twitching.

When did you get your 400, and what was it?

A Yellow-throated Vireo at Kenidjack in Cornwall in 1990. I was on Scilly in September with Pete Fraser and we came off Scilly for the day for that bird.

I found myself on a kind of 'academic treadmill'. I had the chance to do a post-doc at Sussex with Peter Slater on birdsong, which had a lot of scientific credibility associated with it. Peter interviewed me and said that the post-doc started in August, which was six weeks away, and I could have it if I was written up. So I worked 24/7 to get written up in time, which I did. For the post-doc I worked on Chaffinch song in the field in Stanmer Wood, looking at the functional aspects of birdsong, and whether repertoire size has an impact on territory size or breeding success. On the side I also did a piece of work in the lab on Zebra Finch mate choice through song and, as with lots of things, that was the bit of research that caught the public's attention. It was in *New Scientist*, and on the back of that Pat Bateson from Cambridge invited me there to do a bit of work.

It was a bit of an eye-opener for me, but by then I'd clearly decided that I wanted to do nature conservation. Peter Slater wanted me to stay at Sussex, but towards the end of my two-year fellowship (1980–82) I had decided I was going to leave academia and work for the Nature Conservancy Council (NCC). I realised I would take a big hit in salary and it would also be uncertain, as there was a lot of competition for conservation jobs with the NCC at the time. I went to Edinburgh to be interviewed by Peter Pitkin and Nigel Easterbee to work in the Scottish Field Unit doing upland bird surveys, and I got the job. I'll always remember the interview because Nigel Easterbee started by asking me what I knew about Merlins, and after I had talked for about three minutes Peter was kind enough to suggest that they give me a job. Peter has remained my close friend ever since. He lives in Edinburgh, and will have retired from Scottish Natural Heritage by the time this book is published. I did about three years' work for the NCC on short-term contracts on upland bird survey, data analysis and other stuff for the Chief Scientist's team. I lived in Edinburgh, and also rented a cottage for a season in Wensleydale to do the Yorkshire Dales, and then latterly I went on the short-term contract staff at Huntingdon when the NCC Chief Scientist's team was based there.

My aim was to get a permanent job with NCC, which became a mission as NCC was 'it' as far as conservation was concerned. I wasn't successful on a couple of occasions, then in 1984 there were four Conservation Officer jobs advertised. I applied (there were 3,000 applicants, which was the nature of the competition at the time), but I got the job as I'd been around, and people knew who I was, and so I became Conservation Officer for Hertfordshire. I stayed at NCC for three years till 1987. We opened a new office in Letchworth and I lived in the Chess Valley between Rickmansworth and Chesham. It was a really nice part of the countryside – I remember at the time hearing Quail from my bedroom window and seeing Osprey and Marsh Harrier flying over.

Not long into this role, my partner's organisation moved to Cardiff, which caused a bit of a problem, but I managed to get a secondment to the Department of the Environment in Bristol to become Chief Wildlife Inspector between 1987 and 1991. The work entailed running a panel of inspectors who checked out keepers of birds of prey – for example, the close-ringing of captive-bred chicks and doing unannounced inspections to find out what birds those guys had, because the level of bird of prey crime then was quite high. Other work included advising Customs and Excise on international trade in endangered species.

That's interesting. Did you see the Scops Owl in Hampshire in 1979? There are rumours that it was an escapee.

That was when I was at Teesside. We came down to Hampshire in the night and the bird was calling – there were five of us in the car. Three saw it and two didn't, one of whom was me. Martin Blick, who was driving, said we didn't need to get back, so let's go to the New Forest, do a bit of birding and come back in the afternoon. So we arrived at the Green in Dummer in the afternoon and parked under the tree. There was this guy sitting on a bench, who asked if we were looking for the Scops Owl. We said we were. He said his name was Paul Sterry and he had discovered where it was roosting. He took us into this garden and the bird was roosting in a pine tree only ten feet above the ground, so we got fantastic photos. That shed some light on the story of them trying to get it back into captivity, which I don't think is true, as they could easily have got it from where it was roosting and this was right at the beginning of its stay.

In 1991 NCC was broken up by Nicholas Ridley and Malcolm Rifkind, and English Nature was created. I returned from secondment to English Nature as Head of Science in South Region, based at Foxhold House in Newbury.

I started to do a range of jobs in English Nature in Peterborough where I met my wife, and mother of my children, Susannah, who was a consultant with Price Waterhouse. Her business influence gave me the confidence to broaden my science expertise and include business elements in my approach to conservation. I did a piece of work for one of the directors about products and services in English Nature, which singled me out as someone willing to do stuff and get on.

This gave my career a boost and I then turned my hand to a variety of roles within English Nature. I set up and ran the European project, became Corporate Manager and wrote the organisational strategy, then became a General Manager. That's when I met Barbara Young, who had been Chief

Executive of the RSPB and became Chair of English Nature. At the time I had the Communications function in my portfolio as the General Manager, together with a clutch of teams I had to manage and visit. Two of mine were Dorset and Hampshire. I was visiting Dorset and she was visiting Hampshire and we got on the same train in London and chatted. She got out at Lyndhurst and said, 'Save me a place on the train on the way back,' and got on at Lyndhurst on the way back. She said, 'That was an interesting visit. There's a big public inquiry coming up – there are proposals to build a new container port at Dibden Bay in Southampton Water, and we need some senior leadership. You need to do it, Andy.' That made my career, because I set up a team, spent about £750,000 on staff and consultancy, and we had David Tyldesley Associates – the best in the business – doing the planning stuff. Our advocate was Graham Machin, a brilliant barrister from Nottingham. It was a great experience, and on top of all of that we won the public inquiry; the biggest infrastructure turn-down by government on environmental grounds ever. It was a £650 million development and it cost Associated British Ports (ABP) £100 million to lose it. One of the things it taught me was the importance of partnerships to override particular issues. The CEO of ABP phoned me a couple of weeks after the judgement and said he'd like me to come to their board and tell them where they went wrong. I said I didn't want to do it, but he said he wanted me to tell them the truth, and I did. The reason they lost was because their legal team ran the show and weren't properly managed, and they chose the wrong route. They could have gone ahead with the development, with nature conservation built in, had they chosen to do so.

Following that success, I became Director of Protected Areas and that was the best job ever for me at the time. I thought it was wonderful, and that's where I learned how to do a lot of change management. I had about 400 staff and £20 million to spend on protected areas management and we had some very big successes.

The chairman following Barbara Young was Sir Martin Doughty. He was a strong Labour man, and he was passionate about access to the countryside. His father had been on the Kinder Scout Trespass. I was once again working direct with the Chief Executive. Martin and I got on very well, but very sadly he died of cancer in 2009. I went to his memorial in the local church hall in New Mills in Derbyshire where he lived, and Roy Hattersley spoke. He said, 'Many of you will have seen the piece I wrote in the *Guardian* about Martin Doughty the professional, but I was Martin Doughty's friend and I'm going to talk to you today about Martin Doughty as my friend, but of course Martin Doughty the friend and Martin Doughty the professional are the same person.' Inspiring stuff.

In 2005, English Nature amalgamated with the Countryside Agency and the Rural Development Service (the part of Defra that delivered agri-environment funds to farmers) became Natural England, and I was a champion of the change programme. Out of the fifteen directors in those three organisations, I was the only one to become a director in Natural England – Director of Science, Evidence and Policy.

It was not a management culture that suited me, driven by the new Chief Executive and with two other directors who didn't have a scientific or nature conservation background. It seemed as though all I'd worked for to date was being disregarded. In the end it became too much for me. I was working too hard – seventy hours a week – and didn't feel as though I was achieving anything worthwhile. It was a very challenging time, and so I quit.

Having had a difficult end to a brilliant career with the government agencies, were you at all tempted to move away and do some consultancy work completely outside of this?

I was not in the best of health – I'd left my job with a stress-related illness which was not like me at all. It was summer 2006, very hot and sunny, and I took time out to recover for three months and think about things. I decided to set up my own consultancy, and called it Ferrypath Consulting. I did some strategic environmental consulting for Anglian Water Group, the Cambridge Programme for Sustainable Leadership and the wildlife travel industry. I quite enjoyed it, but was used to working in organisations so wasn't entirely comfortable with the change, but I thought I'd made my contribution and the consulting would do.

Do you mean you need to be driven by people, or need to be driving people?

A bit of both. I very much like the team feeling – a well-directed, well-managed organisation can achieve an awful lot, and my experience had been about building and leading teams that did great things. It's not about you as an individual at all; it's about what the team can do collectively.

Looking back, do you think you should have left earlier and not put yourself through the stress?

I could have, but I was only there for five months and didn't know I was going to put myself through that stress.

So I set up Ferrypath Consulting, and in much the same way as when Mark Avery left the RSPB, people like me made sure we always invited him to things, so people then invited me to things. One of the first events I went to was at the Royal Society and the first person I saw was Jeremy Greenwood, who said he was leaving the BTO. I said I thought that would be an interesting job and we arranged that I would go and see him a month or so later. I'll always remember that day. It was 24 October 2006, the day before my birthday. I drove to Thetford, had a morning with Jeremy talking about the job, and in the afternoon I went and saw the Red-flanked Bluetail at Thorpeness – and because of its significance I still have a photo of that bird on the wall in my office!

I'd always thought that running the BTO would be a great job. I'd been a member in my teens and while I was a student, then – like perhaps many people – my membership lapsed in 'mid-life', so I thought I'd better join again straight away! I was very fortunate to get the job, and I'm the first BTO director not to be a practising academic scientist.

Could we talk about the BTO, and the role it plays in bringing birders and science together: what do you want to achieve?

I've been at the BTO for five years. When I went there I thought that one of the remarkable things about it was that it was a best-kept secret: they did great science but nobody knew about it. There is of course a traditional core of BTO membership, and a traditional set of partners, who have always known about BTO science, but in terms of public profile, the specialist profile within the nature and conservation community, and the appeal to birders, it was far too small, so I thought there was plenty of opportunity to make that better. I'd always believed that the BTO was a conservation organisation, in that its product was hard scientific evidence, and that informed nature conservation in many different ways. There are still people, like Mark Avery, who state that the BTO isn't a conservation organisation, which I disagree with. We're not a lobbying organisation, and we don't achieve our aims by telling people what they have to do, but people can't make decisions on many environmental matters without much of the evidence that is provided by the BTO. So I'm passionate about making the BTO count in all sorts of different ways.

So you're not actually kicking the door down, but you're showing someone where the door is?

Definitely. We're not the only people doing that, but I think we do it very, very well. I think the quality of the BTO's science is phenomenal.

The other thing we haven't told a story very well about is the interplay between – to use an American term – citizen scientists (or volunteers, as we call them) out there and what professional scientists can do as a result. Those two groups work together very well in the BTO, but until now, we have not made the benefits of that clear. The platform to do so was there, but we haven't been very overt about the end product. The volunteers have collated long-term data sets over the years (they produce their own products such as the Wild Bird Indicator and Farmland Bird Index, which is a classy piece of work and goes to the government at the highest level), and these data sets are used by professional scientists to undertake applied research: James Pearce-Higgins's work on climate change, Gavin Siriwardena's work on agriculture, Niall Burton's work on modelling monitoring methods on offshore wind farms are all phenomenal applied science using what volunteers have collected, which I think really adds to the value of what volunteers have provided over the years. It's not just about the data sets fuelling an indicator; it's about them being a working ground for science ten or twenty years hence. Being able to tell that story, and interest the whole birding community – from twitchers to serious patch workers, to people who are just interested in birds in their garden or would like to do something small in terms of identifying or knowing about their birds – is my mission at the moment.

What is the BTO going to look like in ten or twenty years' time?

In ten years' time I'd like our membership to be roughly double what it is now, I'd like our volunteer network to be double what it is now, so we will have 40,000 members and 100,000 active volunteers. I'd like the demographic to be different too. We're doing lots of work in developing Birdtrack as the premier bird recording portal for members of the public: its products and services for birders, like league tables, its community feel, information on where birds are being seen in your local area, and merging bird information services with a platform into which you can enter data for us to work with. So, we're giving birders things that they want. We've done lots of analysis, so for instance we need to know why people who put their lists on Bubo are not the same people who put their data into Birdtrack. Now that we're a bit better known, people assume that birders everywhere like the BTO, are familiar with us and members. It's not yet the case, but in ten years' time it will be.

One of the problems, surely, is that your staff are very well-qualified scientists, and although they watch birds, they're not really birders first and foremost?

I would disagree with that. I think we've got very well-qualified scientists who don't watch birds much, and some who do, like Simon Gillings! But we also have a range of other scientific and less scientific staff who are big birders – Andy Musgrove, Nick Moran, Dawn Balmer – so not only are they good birders, they're evangelical about birding and the BTO together. If you look at Ebird in America, you'll see it is an amazing way to record your bird sightings in North America. One of the key developments is the evangelical nature of people like Chris Wood, who are scattered all over the States and work with the equivalent of our county bird clubs and county bird recorders to get them on side with Ebird. Increasingly, I think people like Andy, Nick and Dawn are doing something similar.

Why hasn't this been achieved before?

Because bird science, rather than birdwatching, was seen as the thing at the BTO. Since I've got there we've been able to raise the profile of birdwatching as well as the science. I think the fear was that if BTO did both, that would dumb down the quality of the science and the BTO would become a less serious organisation. Well, I think we need to be more accessible to a wider range of people, and we can do that without the slightest hint of dumbing down the science, which is of a higher quality than it's ever been. Also it's more noticeable. This business about not lobbying government and not bashing people about the head about what they should do is fine, and I maintain that line ferociously, but we can go right up to that line for policy relevance with the work that we do. So we're getting money from the government for a lot of the international, marine and agricultural climate change stuff, because it's highly policy-relevant for them.

The word 'ornithology' is probably important to the organisations who are going to fund you centrally, such as Defra, but I feel that it's losing its impact with the people who might join. Where do you see the word 'ornithology' in ten years?

Disappearing: well, I'm sure it won't disappear completely, but you will have noticed that we're not called the British Trust for Ornithology any more, we're called BTO.

How about using the word 'bird' somewhere in the name? Will it still be called the BTO in ten years' time?

I think not, in ten years' time. The marketing change that we made a couple of years ago was massive, and it went very well, considering that people could have said we'd thrown away a lot of the BTO's tradition. We worked very hard to make sure we weren't doing that, but we did want to present ourselves differently, and I think it's worked very well.

Tell me about the logo, changing it, and why you waited till 2010. You instinctively knew it had to change?

Yes, I did. I'm a firm believer – and my experience at Natural England helped to confirm this – that new chief executives walk into organisations and want to change everything. Sometimes it works, sometimes it doesn't, and for the BTO that would not have worked. We have quite a long-standing group of staff who are among the top scientists; we have a traditional membership who would be quite upset if we changed things too dramatically, too quickly; and we needed to think about things a bit before taking action. Also, the logo wasn't the main thing, I wanted to sort out the strategy first, and once that was done, we had a platform for those other changes to take place.

Many people said at the time they didn't like the new logo. What do they say now?

I actually only remember hearing about twenty people say they didn't like it. I remember someone walking out of the presentation I gave at Swanwick when they heard how much the whole thing had cost. I still hear a few people say they don't like it – last week on Scilly I met a friend who said he didn't particularly like it, but he liked the direction in which the organisation was heading. But most people just accept that's how the BTO is now. Which is a great thing, because it works, and it's opened up opportunities for us to be much more accessible and inclusive with a much wider range of players. That's why our membership is growing: we had another 8–10% growth this year.

I joined when I was fifteen – why do most people wait to join until they've got kids aged fifteen?

Because they don't see it as relevant to them at their time of life, and I think that's part of the exclusive, slightly elitist, approach to date. Things like the apps are geared to younger people. Having an Android app and putting Tweets in presentations, that's how people communicate these days. Of course, the BTO's social media presence is really good: we have over 10,000 followers on Twitter (I have over 1,300 followers myself), and an equivalent number on Facebook.

You don't have any young people active on BTO committees, so how can you expect to know how younger people feel and think? If everyone you see is your parents' age, why would you join?

You're right, and so the next thing we want to do is get a youth council going. The last session at the recent BTO conference was about young people. We had three students speaking who collectively have done more than 1,000 nest records this year, as well as doing ringing.

Would you like to be admired or liked by the people who work for you?

One of my faults is that I want to be liked too much. Not admired; I just want to be part of a team. Leadership is about being clear, straightforward, and visionary about leading the team, but it's also about standing next to, and being with, the people you're leading. It's a difficult trick to pull off, but I hope I bring my experience of some success at English Nature into being a good leader at BTO. It's another reason for not changing everything straight away – the kind of leader who does that is out in front all the time, and that's not what I am.

How long do you think you'll stay with the BTO?

I really don't know. Staying in the top position for more than ten or twelve years may be too long, and I've done five.

You've obviously met a lot of impressive people in your career. Imagine you could meet a couple more people, alive or dead – who would you choose?

One would be Aung San Suu Kyi, because she's the kind of leader I want to be, combining humility and personal power. She's put up with a lot and still has presence. She has personal power, not positional power, having essentially been in prison for twenty-five years, rather like Nelson Mandela. If I could ask her one question, it would be what advice she has for me as a leader.

I've met David Attenborough, but I'd like to have a proper conversation with him as he's been the one person in my lifetime who's been an amazing communicator. Once he opens his mouth, you want to hear what he has to say.

Also Max Nicholson, the founder of the BTO and Chief Executive of the Nature Conservancy in the 1950s. He laid the ground for all the things I've been able to do in my professional life.

What's the best birding place you've been to?

Bhutan, in the Eastern Himalayas – pristine forest. I was there in 1986 with my partner at the time who was doing a beekeeping job, and the country wasn't open to visitors then. We found Black-necked Cranes, which was significant, and I wrote a paper about that; and saw the first Rufous-necked Hornbills in the Himalayas by a Westerner for a century. I led Naturetrek's first trip there in 1990 jointly with the Himalayan botanist Tony Schilling.

Where's the worst place you've been birding?

I actually can't think of a worst place! The presence of birds makes everywhere a better place.

What's the best bird you've ever seen?

Satyr Tragopan in Bhutan.

What's your favourite bird family?

I like *Phylloscopus* warblers – I've seen them here as they're in our daily lives, such as Chiffchaff, Willow Warbler, Wood Warbler and rarer ones like Pallas's Warbler. Also there are so many in the Himalayas.

What's your most wanted bird in Britain that you've not seen?

Pechora Pipit – I haven't seen that anywhere in the world.

If you could see one more bird in the world that you've never seen before, what would you choose?

Picathartes, because it's so different from anything else, and it would take me to a part of the world I've never been before.

If you could have an airline ticket to somewhere in the world you haven't been, where would you go?

The southern part of South America: Tierra del Fuego.

Imagine you're on a desert island and you could have five pieces of music. What would you choose?

Some Handel arias sung by a counter-tenor; something to remind me of my kids now – something like Florence and the Machine; Bach's *Cello Suites*; some Britten opera to remind me of East Anglia, perhaps *Peter Grimes*; and something from my growing-up period, such as Pink Floyd or Dire Straits.

What's your favourite film?

The Shawshank Redemption.

Favourite TV programme?

Spooks.

A bird book to take to the desert island?

Wisdom of Birds by Tim Birkhead.

And a non-bird book?

Ted Hughes's *Collected Poetry.*

MIKE CLARKE

Dr Mike Clarke is the Chief Executive of the RSPB. He was born in the 1960s.

INTERVIEWED BY KEITH BETTON

What is your first birding memory?

Probably around the age of eleven going up to the nearest piece of woodland on the edge of Gravesend where I lived: a really nice mix of woodland where I got my head round the various woodland birds in my *Observer's Book of Birds*. I'd always been fascinated by wildlife, and this was getting to the point of more purposeful activity. The woodland was a great place for Hawfinch – not that I saw those on my first birding trip! – but this piece of woodland has recently been taken out by the Channel Tunnel rail link. That part of Kent is a very urbanised environment.

Did anyone influence you at this time?

No, it was self-generated, and probably the biggest influence was the *AA Reader's Digest Book of Birds*. I could recite by heart almost every page of that book (I couldn't do it now!). It was a mine of information and opened up my world. I can remember the pictures even now.

What were your first binoculars like?

I think they were probably my grandfather's – an Eastern European make of some description. Having discovered birdwatching and the AA Reader's Digest book, I joined the YOC at the age of twelve. I got the information through, I was getting so excited by all the stuff I was learning and the places I could go to – I remember telling this friend of mine all about it at school registration, but he was totally unimpressed! This made me realise that this was a world you only discover if you make the effort, but if you do make the effort you get such a lot out of it: wonderful friends, places and fantastic birds.

So having joined the YOC in 1972 or so, almost immediately a local group was formed by Ann Scott (before she married Bob), who later worked for

Richard Porter when he set up the South-east Regional Office. This started me off going to places and getting to grips with estuary birds and woodland birds, plus a whole range of habitats. Also it connected me with other kids who were just as keen as I was, and that was when it took off for me. We went on a few coach trips, but a group of us did our own thing – we lived with the north Kent marshes on one side and some fantastic woodland on the other, so we discovered this world together, which is quite an intense memory for me. You could see Shorne and Cliffe Marshes from my school and once I'd connected with these friends I was there literally every weekend. I worked out that, in the Lower Sixth, I was at Cliffe once a week every week for a year. The second thing, when I was a bit older, was going on various volunteer reserve working parties (Northward Hill, Dungeness and various other nature reserves), which was great fun and also created a sense that, by doing things together, we could really make a difference.

Within three years I was using Sandwich Bay Bird Observatory quite a lot and by the end of the sixth form I was doing voluntary warden work there, keeping the place ticking over. I was Little Tern warden for the Kent Wildlife Trust and also did voluntary wardening on Skomer, which was fantastic. By the time I'd done my O-Levels a group of us had been independently to Skomer and Skokholm, and cycling round the Loire Valley at the wrong time of year looking for all sorts of birds, but seeing things like Black Woodpecker.

By the time I started my A-Levels, I was pretty sure I wanted to get into ecology and conservation, driven by the experiences I'd had birding, and some of the stuff I'd been doing on reserves, as well as the work that people like Dick Potts were doing on Grey Partridge. That type of field ecology was the kind of work I wanted to do, which is why I studied zoology at Oxford. Lots of interesting people came to lecture, and Nikolaas 'Niko' Tinbergen was there at the time. Nick Davies was my tutor, along with Chris Perrins and John Krebs.

You haven't mentioned much about birding outside Kent.

At school you're limited pretty much to bike, train and bus, so I'd gone on one or two trips to Minsmere where I saw Bearded Tit and Marsh Harrier, which was a big deal as there were only seven pairs in the UK at the time, and I vividly remember when they were spreading and first colonised north Kent. I went on a YOC holiday to west Wales based at Borth Youth Hostel, so I have early memories of watching (from the hide at Ynys-hir) Cormorants grappling with huge eels on the salt marsh, and looking for things like Chough and Red Kite. I also hitched down to Wales on my own, and went over to France, but it was really only when I got to university that I was able to think about starting to travel more widely.

What was your first pair of proper binoculars?

About the age of fourteen, a pair of Chinon 8×40, which I still have on the windowsill. I then upgraded to a second-hand pair of Zeiss Jenoptem 10×50 at university, kept those for quite a while and then got a pair of Zeiss Dialyt, which I'd only ever dreamed of. I treated myself to those when I joined the RSPB staff, and I still use them, but I'm going to upgrade my scope and binoculars to RSPB ones (the top end of our range is now something I can be proud of!). I got my first telescope after I left university – a budget-issue Opticron with a couple of zoom eyepieces.

What bird books do you remember buying or using a lot in those days?

There was an explosion of information coming out at the time (but nothing on the scale of the internet, of course) in terms of field guides. I was hungry for identification books and I used to subscribe to *British Birds* magazine, which had loads of identification papers in, and then more of the monograph type books, such as Harrison's *Seabirds*, began to be published. One book I vividly remember was the *BTO Guide to the Identification and Ageing of Holarctic Waders*. If you read through that, there was quite a bit you could use in the field as well as ringing, such as identifying races of Dunlin, which you could work through if you got good views. I remember being glued to that and also the great book on shorebirds by Tony Prater, John Marchant and Peter Hayman (*Shorebirds – An Identification Guide to the Waders of the World*) – it dropped through the letterbox and I remember sitting transfixed on the bottom step going through every page. When I was down at Sandwich Bay, which I was a lot for a few years, there was a copy of Witherby's *The Handbook of British Birds*, which we penniless schoolkids always aspired to having but could never get, then the *Popular Handbook* came out so I got that. But when they started producing *Birds of the Western Palearctic* I just had to have it and got it for Christmas and birthday presents. Then, unfortunately in many respects (like shelf space!), Del Hoyo's *The Handbook of the Birds of the World* came along after that so my family has had a fairly easy run of present-buying!

Did you photograph or make sound recordings of birds?

I took a few photos, but I didn't have a telephoto lens, which was expensive and I didn't have the money. In fact, I may still have a photo of an albino Manx

Shearwater which I took on a Kodak Instamatic. I didn't make sound recordings.

What about family holidays?

We weren't flush with money. I went with Mum and Dad to the Lake District and Norfolk, then to south-west France, but once I got into my mid-teens I travelled independently, in this country initially, then to places like the Loire Valley after my O-Levels.

It doesn't sound as if your parents played much of a role in your birdwatching; it was all pretty much self-started.

That's true. My parents were hugely supportive but no one had an active interest.

Did you keep a notebook?

Yes, and I still have it. I did mainly lists of what I'd seen in the day but I didn't do many sketches – a few field ones, but that's all. I appeared in the Kent Bird Report with a Dipper flying off the sea at Sandwich Bay and landing in the sea, but I had no amazing rarities. I did see a record number of twenty-six Jack Snipe in Kent, but I don't think I can claim any stunning rarity.

When did you decide you wanted to work in conservation?

When I was doing my A-Levels I knew I wanted to work in ecology and conservation but I didn't have a clear enough idea of what jobs were available at that time. Then when I was doing my degree I was even keener to do so. But when I finished in 1981 it wasn't a great time to be looking for jobs, as it was the year of public spending cuts and by that time I knew I wanted to work in nature conservation. I had a place and a grant for the MSc Conservation course at University College in London, but I knew someone who'd done it the year ahead of me, who was still looking for a job and had decided to do a PhD. So I thought I might as well apply for PhDs as well. There were two that year sponsored by the Nature Conservancy Council – one was with Derek Langslow on shorebirds on the north-east coast and the other, not related to birds, was with Colin Tubbs on valley bogs in the New Forest. I applied for both and was shortlisted down to the last two for the shorebirds one, but the best person won (!), and that was Rowena Langston. I got the one in Hampshire. I worked for the NCC for eight years in Lyndhurst and did my PhD during that time, so I started off as a university student but did a few wide-ranging contracts with NCC, which was very good background and broadened my interests. At the time, I was living in the Southampton area and commuting up to Lyndhurst.

Were you also birding in the Forest at that time?

My work was mainly plant and vertebrate ecology, with a lot of lab-based work, but I was doing a lot of field surveying and site survey for NCC. I got to know Dartford Warblers very well, as well as all the other special birds of the Forest. This was at a time when the government had modernised its Site of Special Scientific Interest (SSSI) legislation which required every SSSI to be re-surveyed, so I did a lot of woodland, meadow and heathland surveys.

Did you still go birding at the weekends?

My PhD work was fairly full-on, and I was writing up while working, so weekends didn't really exist. Also, I was married by this time – I met my wife at Oxford.

Would you say that during this time you were a scientist rather than a birder?

In my day job, yes, but on holidays we'd go off birding – the Seychelles, Nepal, India, Thailand, North America, Caribbean, Tanzania. I didn't complete my PhD during the time I was at the university, so I was finishing that off while holding down a full-time job.

When did you join the staff of the RSPB?

January 1988 from NCC. I'd been looking for particular career paths in NCC which weren't available. I was working for Lynne Farrell who wrote the original Red Data List for British plants. As well as rare plants, she was also responsible for heathlands and wanted the heathlands job to be set up as a separate role, which I was keen to do. But after a couple of budget rounds of it not getting through, I thought I'd better look elsewhere and the RSPB was advertising in the *New Scientist*. I got the job as Conservation Officer in south-east England – at the time I was recruited, RSPB was just setting up specialised regional roles. There were five or six of us in the regional office, and I was working for Andy Bunten, who had taken over from Tony Prater. I had that job for three or four years, then in about 1991 I became what is now known as Regional Director. When I started as Conservation Officer, I went back to look at the issues in the places where I birded when I was growing up. They were very vulnerable at the time, under huge development pressures, and it felt like a large part of my job was preventing this, by putting the government on the spot and making it impossible for them to give the go-ahead for development.

When you got the Regional Director job, did you envisage yourself ever being at the RSPB headquarters at Sandy?

I didn't really think about it; I didn't really have a predetermined career plan. I'm not sure I believe people who say they do!

When our Director of Regional Operations was moved on in 1998 I decided to apply for his job, and got it.

You are the RSPB's Chief Executive – what's it like?

When I took on this job in 2010, the *2010 United Nations Climate Change Conference* held in Cancun, Mexico, had just failed, and the 10th Conference of Parties (COP) to the Convention on Biological Diversity (CBD) (held in October in Nagoya, Japan) faced up to the fact that biodiversity continued to decline. This was a moment when the world's politicians failed to provide global leadership – when they needed to take the right decisions for people and the planet, but didn't. This was a signal that charities would have to play a bigger role. We have to be the voice for nature, and work with right-thinking people, whoever they are, to ensure governments do the right thing.

Despite the many successes of the RSPB and other conservation organisations, we are winning battles but losing the war. During the time I have been an RSPB member, there are 40 million fewer individual birds now breeding in Britain. The RSPB and twenty-four other organisations produced the State of Nature report in 2013 which provides a compelling analysis, showing some 60% of biodiversity is in decline. Wildflowers and the seed bank are diminishing, as are insects and birds. We are seeing a dismantling of food webs, a reduction in the biomass of wild birds, and a loss of diversity. The problem is systemic and the solutions also need to be systemic. It means that bird conservation – and nature conservation as a whole – has to make itself more relevant to other agendas, such as health, poverty and human development. We need much more collective impact for common cause.

We need to ask ourselves some tough questions and then be prepared to think and act differently. For example, how we can do more for birds and all of biodiversity? How can we significantly grow public support for nature conservation? The challenge is too big for any one organisation – so how can we best collaborate with others? We have a rich history to draw upon and it gives us the springboard for our future development. It is about evolution and the next steps from a proud past.

How we can do more for birds and all of biodiversity?

The RSPB has been on a journey since the Earth Summit in Rio. We have reached a milestone and now have to be ready to move on. By the 1980s, we had become an early adopter of the principle 'think global, act local'. We were on a course. We started by identifying the threats facing birds. But tackling the causes of the problems meant we had to look through the other end of the telescope to find solutions for nature as a whole, because that is how nature works. Birds don't exist in isolation, they are parts of food chains, habitats and flyways. For example, we started by protecting Black-throated Divers and other fantastic birds of the Flow Country, and ended up successfully campaigning on capital gains tax and UK forestry policy. We started tackling the decline of the Skylark and ended up influencing the largest single financial trade instrument in the world, the common agricultural policy (CAP). From the very roots of the RSPB, we started protecting Kittiwakes on seabird cliffs, and ended up involved in fishery management, marine protection and, increasingly, energy policy.

So where are you today?

Today, the RSPB is building a more formidable toolkit and stronger partnerships, which means we are making a major contribution to the conservation of all of nature, not just in the UK but also globally, as part of the world's largest partnership for nature and people. Birds will remain at the forefront of what we do locally and globally – birds are important indicators of the health of the environment, birds are a very visible and tangible part of the natural world wherever we live. But we have to recognise that the natural world is in crisis. I know RSPB can do more – and right now, nature needs us to do more than ever before. This is a natural progression and I want us to build on our strengths.

Where is the RSPB going?

We already do more than it says on the tin. We've been doing a lot of the things that help to save nature for many years. For example, take the fantastic recovery in the Bittern population. We have achieved this by creating new wetland habitats with good water quality, and rebuilding the food chain with increased numbers of invertebrates and fish populations, so that we can see the return of top predators like Herons and Otters as well as Bitterns. And we are taking care of 14,000 animal and plant species on our reserves. So first of all we need to be recognised for what we are doing already.

Second, our plan for how we can do more for birds and the rest of biodiversity includes developing a common evidence base on the state of nature, agreeing species priorities beyond birds which we think we can help, and adapting our toolkit to work harder for all nature; for example, we work with farmers to help them do their bit, for birds like Lapwings. We will adapt the advice we give on wildlife-friendly farming, with others, to cover all nature.

How can you significantly grow public support for nature conservation?

Nature desperately needs more people to care, and to act. In a world where people are bombarded with all kinds of competing messages, we need to ensure that the RSPB's voice is heard so it chimes with the pace of life in the twenty-first century. One of the main barriers we have to overcome is the sense that many people have that it is all too big and hopeless. This is a current challenge for all conservation organisations and many other charities. We have to keep up with change, and that means we have to question what makes us most effective in the public eye. We have revisited how we project the message about our cause to the public. We believe that we need to reach out beyond the core base for nature conservation, and raise the popular profile of nature and grow our number of supporters. We hope that this will help the sector grow in practical, moral and financial support, and in turn help save nature.

We have embarked on a major public engagement campaign to help change public perceptions about both nature conservation and the RSPB. We need to inspire people to take action in their own lives, to inspire action in communities, in the wider countryside and globally.

That's why our campaign talks about 'Giving Nature a Home' and encourages people to take individual action, in their own lives and communities, as well as supporting our work to give nature a home on our reserves, in landscapes and across flyways.

But surely this challenge is too big for any one organisation – so how can you best collaborate with others?

I agree with you. These challenges are definitely too big for any one organisation. In fact, they're too big for any one sector. We need politicians to act more in the long-term public interest than they do now, and businesses to be more enlightened about their long-term interest too. This means that civil society is going to have to set the agenda and pace much more, with charities like mine working alongside faith groups, the arts and many other causes. We can't save birds on our own, and we can't save nature on our own. Last year, as I mentioned earlier, the RSPB was involved in bringing out a vital new report on the State of Nature in the UK. Unless we know what the state of nature is in the present, we can't protect it in the future. And if we leave it to the government, we won't know what we've got until it's gone.

But it's bigger than just nature, isn't it?

Definitely, and in 2012 the RSPB launched a major new initiative with the University of Essex to provide hard evidence on people's connection with nature. And we are part of the Wild Network, a new partnership with the National Trust, Sport England and the National Health Service and many others, a campaign to reclaim nature as part of childhood.

The RSPB is seeing a picture that is much bigger than just birds. How big is the picture?

Well, for the first time in human history, more than half of humanity lives in cities. As a species we are becoming more and more disconnected from nature, with busier and more urbanised lifestyles. This is a strategic threat to nature itself. We need to reconnect people with nature. The importance of nature needs to be recognised by society as a whole. Quite simply, there is no life without wildlife. Today's children will be making tomorrow's decisions, and the challenges they face for people and the planet will be the biggest in history.

As conservationists, we have to open the minds of the next generation to the significance of nature. The RSPB has a long history of winning hearts as well as minds. But, now, it is urgent that we build a much bigger community of people with a deep and lasting commitment to saving nature – not just now, but in ten years' time.

The RSPB currently has just over 220 reserves. If it has, say, 1.5 million members in ten years' time, will there be 350 reserves or nearer to 280?

Looking ahead, I think there's always going to be a trade-off between new reserves and bigger reserves. If you look at this from a birding point of view, it would probably be good for the RSPB to have bigger reserves because some of those would bring in species that we don't currently have because of the range of conditions that they need. Also Sir John Lawton, a very significant figure in the bird world and an ecologist, headed up a major review a few years back which shows all the scientific understanding says bigger, more joined-up, well-managed landscapes are needed to ensure that the habitats remain in the future.

What about the RSPB's role internationally?

We have worked collaboratively around the world since at least 1904, and this will continue. There is a growing emphasis on co-ordinated action for our flyway and the oceans, as well as global themes, tropical forests and the threat from invasive species. Our commitment to BirdLife International is undiminished. We are currently supporting more than two dozen partners in other countries, funding global work on seabirds and important areas for birds and biodiversity. The RSPB has also been raising the profile of the UK overseas territories, and we have a growing programme of action for their biodiversity, which constitutes 95% of the UK's global responsibility for nature conservation.

Where's the money going to come from for the RSPB going forward – do you see that changing?

The support from our members is vital – morally, practically through volunteering and campaigns, as well as financially. We aim for much of our funding to continue to come from individual supporters, as it does now, because it gives us a very high degree of independence from the government, which allows us to perform the important role of challenging the government and holding it to account.

What do you think you'll be doing in ten years' time?

I have no idea. Right now, I'm very focused on ensuring that the RSPB is able to adapt to the scale of change under way, from the growing global environmental challenges to the digital revolution. That's quite enough to think about!

Do you still go birding these days?

I tend to spend weekends (when I'm not working) with my family, but I'm lucky in where we live as there are Swallows, Buzzards and three species of owl breeding next to our house, along with Red Kite, Hobby and Kingfisher. Our holidays are pretty much birds and wildlife-oriented. We are planning for when both our children have left for university, and I am looking forward to having the option of holidays outside July and August!

If you hadn't gone down this career path, what would you have done?

Before I got into birding as a child I thought about being a doctor, but don't think I'd have got the necessary grades. I think a lot of it's down to educational opportunity, so there's no question I'd have gone down the same route, and whether I was RSPB Chief Executive or a warden on a local nature reserve I'd have been doing something linked to conservation.

After birding, what would have been your next best hobby at school?

Some aspect of wildlife – butterflies, plants. I did all sort of sports at school and university, especially rugby and hockey, but they were things that just filled time!

As a birder, and working in conservation, who have you looked up to and admired?

Colin Tubbs, who is unquestionably the single biggest figure in the sense of being a birder and field naturalist and teaching a lot about nature conservation. I learned a lot from a range of people at bird observatories, such as Dennis Batchelor. He helped run Sandwich Bay Bird Observatory and was a bit of a guru when I was a young teenager.

Where's the best place you've been to birding?

The best places are the ones with some meaning to me personally. So although it may not sound that spectacular, in my late twenties I went to Varangerfjord. I was driving with the sunroof open and Temminck's Stints were calling and display flying just above my head. I vividly remember reading about it in John Gooders's book *Where to Watch Birds in Britain and Europe*, and having to imagine it as a boy.

What was your worst day's birding?

When you get a north-westerly wind, seabirds and other waterfowl funnel up the Thames Estuary, and I can remember loads of Brent Geese going past Cliffe. After that, it got bitterly cold, and we decided to leave. Just after we left, a Leach's Petrel went past, which I hadn't seen.

Do you have a favourite bird family?

Waders – I think the Dunlin is a very underestimated bird.

If I could give you a free plane ticket to go birding for a day somewhere, where would you go?

There are three places. I've never seen an Albatross, and I'd love to see a Wandering Albatross in particular, as its wingspan is exactly the width of my bedroom when I was a kid. I'd love to see a Spoon-billed Sandpiper. And I'd love to go and stand on the edge of the ice sheet in the far north-west of Greenland just as it's melting and all the food under the ice becomes accessible, and see millions of Little Auks and Narwhal and all the other teeming Arctic wildlife.

Favourite music?

Quite a range, from folk and blues to classical, but mostly 1960s and 1970s rock and roll – probably Bob Dylan or Joni Mitchell are my favourites.

Favourite film?

Probably *Don't Look Now*, starring Julie Christie.

Favourite TV programme?

TV? That's what my kids watch; I don't have time!

Imagine you're on a desert island. You can have two books, one a bird book and another, a non-bird book. What would you choose?

Seabirds by Peter Harrison, and *Northern Lights* by Philip Pullman.

DEBBIE PAIN

Dr Debbie Pain is the Conservation Director of the Wildfowl and Wetlands Trust (WWT). She was born in the 1960s.

INTERVIEWED BY MARK AVERY

We're sitting in your office at Slimbridge, looking over the Rushy Pen. Have you ever seen any good birds out of your window?

A Black Tern, that was pretty good; that was last year. But I don't spend a lot of time looking out of my window.

Why not?

Because I'm too busy working, trying to save the things that I'm not looking at out of my window! The one thing I always do, though, in winter, when we get a really good Starling roost, and there is a fantastic sunset, is to make a point of stopping for five minutes and watching the starlings coming in. It's one of those wildlife spectaculars.

Tell me when you started birdwatching.

I started when I was about seven. I remember this really clearly. We had a school teacher called Mr Lewis and one day he stopped the class and took us outside and said we had to be really, really quiet and not move. He showed us a small flock of Linnets. And I think it was the combination of being able to stop class work and go outside, seeing these lovely birds for the first time and thinking, 'Wow! they're amazing!', and being told to be so quiet that made it seem like a very special thing to see – which it was. I went home and said 'I want a bird book.'

Can you remember what bird book you got?

You know, I can't. It was probably one of those Collins guides, but I remember that we also had a large bird book with lots of paintings in it. I think it was an AA Reader's Digest book.

With a Tawny Owl on the front?

Yes, that's it! I really loved that book – it was so beautiful. It wasn't the best book to help with bird identification in the field, but it was lovely. From then on I was interested in birds and nature in general. And I think it was mainly thanks to that inspirational teacher.

Would you call yourself a birder?

These labels! I go birdwatching a lot. I spend most of my spare time birdwatching. I go on birdwatching holidays. I read about birds. I work on birds. I'd describe myself first and foremost as a conservationist – but probably people would describe me as a birder. I love all wildlife, all nature. Am I a birder? Yes, I guess so.

Are you any good at butterflies, dragonflies, other nature?

I'm really into butterflies. I travel around Europe a bit looking for butterflies I haven't seen before. But outside Europe, no. I love seeing butterflies in tropical rainforests because many are so stunning, but I don't try to learn to identify all of those – it's just too difficult. In Europe, definitely. I'm very proud of having found a Grass Jewel – Europe's smallest butterfly – when we were in Greece. We also went to a fantastic national park in western Hungary recently because four species of *Maculinea* butterflies – you know, the 'large blues' – were on the wing at the same time in this really small area. So I've travelled quite a bit to see butterflies in Europe.

You're probably one of the world experts on lead and its impacts on wildlife. Tell me why lead is an issue, and how did you get into it?

I got into lead because when I was at school I was good at chemistry and maths. Although I was also pretty good at creative things like art, I was told that if I wanted a decent job I ought to do science. And in those days it wasn't possible to combine science and art. Because (back then at least!), I always used to do what I was told, I did sciences at A-Level and went to university, at Wye College, to do environmental chemistry. I made sure that it was environmental chemistry rather than straight chemistry because I was interested in the environment and conservation.

I was fortunate to get a first-class degree, and after graduating I was offered a range of PhDs on many subjects ranging from cancer research to malaria control, so I asked for advice from my Director of Studies, Dr Clark, because I didn't really know which path to follow. He asked me what I wanted to do in my heart of hearts and I said, 'Something to do with birds – that's what I have always wanted to do.' He suggested I got in touch with Professor Chris Perrins at Oxford who worked on lead poisoning in birds, as that would be a mixture of chemistry and birds. I applied for a Natural Environment Research Council (NERC) award and got the money! It was one of the best pieces of advice I've ever been given.

So that's how I got into lead and birds. My PhD was mostly about the biochemistry of lead poisoning in wildfowl.

What did you discover?

I tried to find a quick and easy method to detect whether birds were lead poisoned – one that could be used in field stations or at least with fairly simple equipment. In the 1980s, doing lead analysis (using spectroscopy) was quite expensive. I looked at a blood enzyme called aminolevulinate dehydratase. The activity of that enzyme is reduced by lead. I got to work with the US Fish and Wildlife Service for six months because they were also interested in this area of work, and I did field work in Chesapeake Bay. It was great, out in the bay trapping ducks and screening them for lead poisoning. And I did the rest of the work in the UK. Yes, we did work out a quick and easy method – it was pretty effective.

Why is lead important?

It's a very toxic metal. That's why we removed lead from petrol and paint and many other things. There was a Royal Commission on Environmental Pollution report on lead in the environment published in 1983, which highlighted the impact of lead on people, including things like lowering IQ in children and many other health effects. It also looked at the impacts of spent lead gunshot on wildfowl and recognised thirty years ago that non-toxic alternative types of gunshot were needed. A shotgun cartridge contains hundreds of small shot, and only a few of these will hit the birds, most falling into the environment. Lead shot remains in the environment for a long time as lead is quite a stable metal under many soil conditions. It does degrade – but generally quite slowly. Wildfowl deliberately ingest grit, which they retain in their gizzards to help grind up food, and lead shot is a similar size to bits of grit and also to seeds that some wildfowl eat, so they pick up lead shot while feeding. Some terrestrial gamebirds do the same thing. They can then become poisoned as the lead is ground down, dissolved by the stomach acids and absorbed into the bird's bloodstream.

Raptors are also affected, although the pathway is a bit different. They can ingest lead shot while scavenging animals that have been killed but not retrieved, or preying upon animals that have been shot but not killed. Lead poisoning is a major cause of illness and death in California Condors – it almost drove the last few in the wild to extinction – though they are exposed to fragments of lead bullets rather than lead shot.

Lead poisoning kills large numbers of wildfowl. In the middle of the last century it was estimated that several million wildfowl a year died from lead poisoning in the USA. And in America they realised that their national bird, the Bald Eagle, was suffering lead poisoning from lead shot ingestion too. So in the mid-1980s they started to ban the use of lead gunshot over certain wetlands, and in 1992 a total ban on hunting wildfowl with lead was introduced across the USA. Probably because of the way that hunting is regulated in the USA, it was incredibly effective, with excellent compliance within just a few years.

In many other countries it is taking a lot longer – although in some European countries things have progressed quite quickly. The Danes, for example, completely banned the use of lead gunshot for all types of shooting, not just wildfowl shooting, in 1996, and a few other countries have taken similar action. Not so in the UK, though. Let's take England as an example because we know something about compliance with the law here, and the law differs a bit between different UK countries. Since 1999, it has been illegal to shoot wildfowl (and Coot and Moorhen) in England anywhere with lead shot, or to use lead shot over many wetlands and the foreshore.

The trouble is, there is little compliance with the law. There was a study carried out ten years ago and another more recently, both of which showed that, of locally sourced wildfowl for sale for human consumption, about 70% were illegally shot with lead – and that hasn't changed in the past ten years.

Let's be clear: that should be 0%?

Yes, that's right. Hunters know that they aren't allowed to use lead. The British Association for Shooting and Conservation (BASC) has been good at providing information on the law and alternative shot types to use on their website, but people don't like to change their habits. Because a hunter might shoot different species in different habitats on the same day – some that can legally be shot using lead and some that can't – then I suppose that people could see it as a bit of a pain to take two types of ammunition – although of course they could choose to shoot everything with non-toxic ammunition. Unfortunately, there has been little enforcement – the police would have to find a shooter actually shooting lead shot and killing a duck with it for the law to be enforced, and much hunting is carried out on private land, so you can see the problem.

Hunters probably think, 'Why should we change?' If no one's going to enforce the law, then nothing much changes. Another problem is that lead poisoning is a bit of an invisible disease. You don't tend to get big die-offs as with some diseases, but rather continual losses of small numbers of birds that are readily removed by predators and scavengers, so people in the countryside rarely see lead-poisoned birds. Hunters probably underestimate the size and scale of the problem.

So it's still a problem in England – it still kills many wildfowl. In Europe even some globally threatened species like White-headed Ducks suffer a lot from the disease but even for more common species like Pochard it may be suppressing their populations. From the research we've done here at WWT, we know that a high proportion of some wildfowl species still have high levels of lead in their blood.

An effective solution is needed. There is currently a Lead Ammunition Group set up to advise Defra and the Food Standards Agency of the risks from lead ammunition to the health of wildlife and people, and possible ways of managing risks. The majority of people on the group are from hunting and shooting interests although conservation, human health and welfare are also represented. I believe that a solution is needed that ensures that toxic lead gunshot is not deposited in places where it is accessible to feeding birds and risks poisoning them or causing environmental contamination. The use of alternative non-toxic ammunition in the place of lead would achieve that. Hunting organisations now have a campaign to increase compliance with the law, which is a really positive step forward – let's hope that compliance is measured and that their campaign works. Even if it does, though, it won't solve the problem for swans, geese and terrestrial birds that feed on agricultural land and other areas where lead can still be used perfectly legally for shooting non-wildfowl species.

And alternative ammunition exists, does it?

Yes. Alternative, non-toxic, ammunition has existed for a very long time. It's publicised on the BASC website and it's in use in many other countries. The most common alternative ammunition is steel, which is of a comparable price to lead.

I eat the odd Pheasant and Wood Pigeon, and they can be shot with lead, and probably have been, so should I be worried?

Lead is poisonous no matter how it is eaten. Personally, I wouldn't knowingly swallow anything containing a high level of lead – it's not a sensible thing to do. There was a 2010 report by the European Food Safety Authority which highlighted the importance of reducing lead intake in the diet, and found that wild shot game contains high lead levels. Although the whole lead shot can be removed from food at the table, the tiny, often invisible, fragments left behind as it passes through the animal can't, so they are of concern. Recent work from the UK Food Standards Agency has shown that eating just one gamebird meal in a week will give you many times more lead than all of the rest of your food and drink put together.

So should you be worried? Well, I suppose that depends on how often you eat game. If it's just a couple of times a year, I wouldn't worry. If it's often, though, then the food safety advice produced in the UK and elsewhere in Europe suggests that you should think twice.

The critical body systems are affected by lead, and it doesn't matter whether you are a duck or a person, really. The nervous system is especially sensitive, and developing organisms are more vulnerable. So we know that increased exposure can have a marked effect on IQ – particularly for developing foetuses and young children. It's not just children, though; adults are also affected, and increased exposure to lead is associated with increased blood pressure and kidney disease.

When I used to come here to Slimbridge as a kid there were lots of those slightly annoying Nene geese wandering around, which WWT had saved from extinction. We thought that Peter Scott had saved them all personally, and WWT has several projects to save endangered species running at the moment.

Nenes annoying? No way!

But you're right – Peter Scott was the man! We're building on that heritage, and WWT is uniquely placed to use conservation breeding, when needed, to bring some of the world's most threatened species back from the brink of extinction. We are using that for Madagascar Pochard, the world's most threatened duck, and the Spoon-billed Sandpiper, the world's most threatened wader.

Although we've talked a lot about lead in this interview, I really don't think that this defines my career – I've spent far more time working on trying to reverse the fortunes of threatened species and to help save important sites and habitats. For example, during my sixteen years at the RSPB I spent a large part of my time working on Asian vultures, which almost became extinct due to a veterinary drug (diclofenac) that was given to sick livestock which, when they died, were scavenged by vultures. I ran the science part of that project for eight years and it was immensely satisfying when the Indian government banned the veterinary sale of the drug – and even more so now that there is some real hope that the populations may increase.

Tell me about the cutest bird in the world – the Spoon-billed Sandpiper.

The bird born with a silver spoon! It's the only bird to my knowledge that hatches with the round part of the spoon but no handle behind it. I think this makes the chick the cutest thing out.

It's a small migratory wader that travels 5,000 miles from one of the most remote and climatically hostile places on earth, Chukotka and Kamchatka in the Russian Far East, through the Yellow Sea and down to South and South-East Asia, where it winters. We believe there are now fewer than 100 pairs left. Throughout the 2000s it was estimated to be declining by 26% a year, so you don't need to be that good at maths to know it's on the verge of extinction.

You can't do captive breeding in Chukotka because it's frozen for most of the year, so we've brought young birds and eggs back here to Slimbridge. We did very detailed feasibility analysis first. That showed that even if the conservation community manages to deal to some extent with the problems with migration and the wintering grounds (mainly loss of inter-tidal feeding habitats and bird trapping), there is still a high chance the species could go extinct, so we thought we should provide the safety net of a captive population. That's not done in isolation from attempts to save the species in the wild, because the last thing we want to do is to end up with just birds in captivity – we will have failed if we do that. So this safety net should ensure that we will have some birds to reintroduce into the wild if needed. If the actions being taken in the wild are successful, we will still probably reintroduce birds to boost the wild population and help it recover more quickly.

This was one of the most challenging, exciting and intrepid expeditions to save a species ever. Kate Humble described the species as 'a James Bond of the bird world' and she was right.

There had been fairly regular surveys in Chukotka from 2000 onwards and Meinypil'gyno was the only site where we knew that about ten pairs remained in one place. To get there you have to fly to Moscow, then fly to Anadyr, the capital of Chukotka province (where Roman Abramovich was the governor once), wait for the fog to clear, then helicopter down to Meinypil'gyno. People have been held up there by the weather for up to a month – our staff were only held up for about two weeks. You need to get all your kit there too.

Thanks to herculean efforts on the part of our staff and those from our partner, Birds Russia, the WWT team got there and collected eggs – and permission to do that only came through a few days before the eggs had to be taken. Then there were problems with finding nests and collecting eggs and then the worries of incubating them successfully for the first time ever and then feeding the chicks in captivity. We had to move the chicks back to Moscow Zoo for quarantine, which was stressful for them and for us. And then we brought them back to Slimbridge, to some specially built aviaries, where they are doing really well.

So you were sitting here in this nice office overlooking the Rushy Pen while your staff were in one of the most godforsaken places on earth dealing with bad weather, permafrost, biting insects, Russian bureaucracy and the stress of raising baby birds of one of the most threatened species on the planet?

Yep! That's about right! And they did a bloody good job too! Sitting here, getting emails, and knowing that there is absolutely nothing that I can do to help, is obviously trivial compared with what they were going through, but I lost many nights sleep worrying about how things were going. I knew it was up to them and their ability and lateral thinking and problem-solving (which is why I knew we had to have the best people), and Nigel Jarrett, our head of conservation breeding, is one of the best there is. We also had massive technical support to call on from experts at the RSPB and the BTO, so we weren't alone.

There aren't many women in this book – and you are one of them. You've said that you are really part of two worlds: you admitted to being a birder (I think you did, anyway) and you are a scientist. When you go birding, you mostly see blokes, don't you?

You know, I'm not sure about that. When you go twitching it's mostly blokes! That's the stamp-collecting type of mentality, the really obsessive thing that more blokes seem to do. I go twitching sometimes, but not very often, and it has to be something really good – and a bird I'm unlikely to come across on holiday – to get me out twitching.

But I think there are more and more people going on birdwatching holidays as couples and families, so that's changing.

What about being a woman in nature conservation? Is it easier or more difficult than for men?

I think it depends on the job you are doing and your circumstances. So I think it can be more difficult and it can be easier. In bird research I think it can be more difficult being a woman, because there aren't that many female ornithologists. Certainly not as managers anyway. But in mammalogy there seem to be more women. It's bizarre!

Ornithological conservation science seems to me to be a bit testosterone-charged, and I did find that a bit tough, because I had to adapt my own behaviour in ways that I wouldn't normally do. I think it depends on the job you are doing. In the job I am doing now, at WWT, there is a very even split of men and women on the management board, and I like the mix. I think men and women bring different things and different approaches to the job. Being a woman can be an advantage and it can be a disadvantage, but you can be sure that when I think it is an advantage I'll use it!

When you said you had to adapt your behaviour …?

I think I often had to be a lot more assertive than I would naturally be. I guess I'm very British; I don't like interrupting and butting in, but sometimes I've had to do that. I think men can be more assertive than women naturally, but maybe it's more about being British and polite rather than being a woman – who knows?

What advice would you give to an even younger woman going into nature conservation now?

And there are a lot of 'even younger' women these days! My advice would be, don't worry about the assertive men, as they're not as good as you! No, I don't mean that; that was totally tongue in cheek. I guess my best advice would be to decide what you want to do and then go for it. Don't let anyone get in your way.

I've certainly never felt discriminated against by the system and I don't think that happens in the conservation organisations I've known.

First pair of binoculars?

They were 8×30 quite chunky German things – Zeiss Jenoptems. They were absolutely brilliant! I had them right through my teens. When I earned my first salary I bought myself a really good pair: they were the first things I bought – Swarovski Habichts.

Where do you go birdwatching?

I live in the Fens and I go birdwatching at my local patches, at places like Fen Drayton Lakes, Woodwalton Fen, Lakenheath Fen and sometimes I take a trip to the north Norfolk coast.

How about holidays?

We usually go abroad for quite long birding and wildlife holidays. We recently went to Malaysia to Fraser's Hill and Taman Negara, which was cracking! The best bird was a Rail-babbler and the best mammal a Malayan Tapir, that we saw visiting a salt lick while we were staying in a mosquito- and rat-infested hide in the forest – it was worth it, though! As well as birding, we do lots of mammal-watching holidays. We tried unsuccessfully for Snow Leopards while camping in the snow in the Himalayas, but we have been more successful at finding wild Giant Pandas, Red Pandas and Pallas's Cat in China recently.

Favourite film?

The films at the top of my list are all French: *Jean de Florette, Manon des Sources, Être et Avoir, The Apartment* and *Chocolat*. If I have to choose just one, it would probably be *Chocolat*. When [my husband] Duncan and I watched it at the cinema he told me afterwards that he couldn't believe the expression on my face as I vicariously experienced three of my favourite things at once: France, chocolate and Johnny Depp!

Favourite piece of music?

'I Think It's Going to Work Out Fine' by Ry Cooder.

Favourite non-bird book?

My favourite books are even more difficult than films as I devour novels (reading at least one a week). One book that sticks in my mind is *The Bone People* by Keri Hulme. Although I read this some twenty years ago, I remember it as a book where the author paints incredible pictures of the complexity of human emotions – it really made me think.

Favourite TV programme?

Probably *Dr Who*. From hiding behind the sofa in mortal fear of the Daleks as a young child to trying very hard – but failing – not to be terrified by the Weeping Angels as a considerably older year-old, *Dr Who* is completely addictive.

Which conservationist would be your hero or heroine? Who would you like to meet?

One of my conservation heroes is David Attenborough. I'm not sure you would call him a conservationist, but he's clearly one of the greatest ever natural history communicators. He has won over a lot of people to the conservation cause simply by being so engaging. I've met him a few times – he's one of our vice presidents at WWT. The first time I met him was at the Birdfair a long time ago, when he was promoting a film on birds of paradise. When I went and shook hands with him my husband, Duncan, said that he had worked in Papua New Guinea for a few months and Attenborough was incredibly humble and charming and said that Duncan must be the real expert. I've met him a few times subsequently and he's always really impressed me.

Tony Juniper is another person I admire – he is a wonderful orator and he's one of those people who lives what he believes. I have great respect for people who do that – it's a difficult thing to do. I have massive respect for Tony Juniper.

My top person would be Rachel Carson – sadly no longer with us. She's the person I would most like to meet. Again, she was a brilliant communicator and her work changed the course of conservation. She was incredibly brave to tackle the pesticide issue – she faced huge opposition, but government policy changed as a result of what she did. DDT was banned and the way we think about agriculture changed as a result of her book, *Silent Spring*. She was an amazing woman. We would also have had some things to talk about as she worked in contaminants, as have I, although she did so at a time when there weren't many women working in that area. She worked for the US Fish and Wildlife Service and I have spent a small amount of time with them too, when I was doing my PhD. Sadly for me, the similarities end there though! She was such an incredible person – a real role model. She wrote *Silent Spring* in the year I was born, so that feels like a bond too. She's the person I would just love to have spoken to and to have learned from.

Favourite bird?

Definitely Kakapo. It's a heavy and slightly smelly (in a nice way!) flightless ground parrot. It's stunningly beautiful – a wonderful mossy green. It has a weird mating system, with males using a system of tracks between depressions or 'bowls' where they inflate like balloons and 'boom' across the mountaintops to attract females. It evolved in New Zealand where there were no predators, and it's just bizarre in so many ways. It seems intelligent and is so engaging that I think it's a mammal in disguise. When I met Duncan I remember he said, 'If you can show me a Kakapo, I'll marry you,' but we got married before I showed him one! And I've had a Kakapo, just the cutest bird ever, perched on my head nibbling my ear – which was amazing. We can't let creatures like that disappear!

KEITH BETTON

Keith Betton worked in the travel business for many years and now is a writer and runs his own public relations consultancy. He was born in the 1960s.

INTERVIEWED BY MARK AVERY

At what age did you start birding?

My first memory of birds is feeding pigeons on Richmond Green in London. I do remember being very interested in the pigeons – I was probably only about three or four – and I have a memory of going on Tennyson Down on the Isle of Wight with my father in 1966 when I was six years old and hearing a Skylark. But my first proper memory of interest in birds was probably from about the age of eight, when I always wanted people to give me birthday cards with birds on them. I liked the ones painted by Basil Ede, and I always asked for bird books, presented in bird wrapping paper. I was obsessed with the images. I didn't go birding as such, but then my parents bought me the *AA Reader's Digest Book of Birds*, the one with the Tawny Owl on the cover, and inside the cover was a leaflet inviting you to join the RSPB. My parents signed me up and I then got *Birds* magazine, which was a bit dull, with information about their next members' weekend, etc. – all far too grown-up for a ten-year-old. So my parents talked to the RSPB, who said, 'You need to get him into the YOC,' which they did. There was a local outing on 15 January 1971 to Staines Reservoir, so I went on that with my father (who wasn't interested in birds). It was led by a man called Brian Shergold, the local YOC leader. I then went on a lot more of these trips led by Brian – without my father – and in my early teens Brian probably influenced me more than anyone else on what birds were, what I was looking at, and so on.

What were your first pair of binoculars?

My grandfather died when I was nine, and my father gave me my grandfather's binoculars. They were a German-made pair of 8×30s, and I used those for about three years. They were out of alignment, so I just simply shut one eye and watched birds with the other eye, then years later I realised it was actually quite good if you could use both eyes and have binoculars that weren't out of alignment! So I moved up to a pair of Swift Belmont, then Swift Audubon, then went on to proper Zeiss, then Leica, then Swarovski, which is what I use now.

So I went around with Brian, then started organising and leading a few walks with him – coach trips to the Kent and Hampshire coast and the New Forest. But the biggest thing for me was getting into watching birds in Bushy Park, my local park in south-west London. When I was about ten I took my kite (a red kite!), which unfortunately crashed into a tree. After that I went back every day to see if it had fallen out of the tree, which it hadn't. My parents told me not to go any further into the Park than that tree, but of course I did, and I cycled a bit further each day till eventually I knew the whole of the Park backwards. It became my main birdwatching location, and in 1973 when I was thirteen, I bumped into an old gentleman who said to me, 'Have you seen some birds? Are you birdwatching?'

I remember thinking perhaps I wouldn't even bother to hang around because I could see he was walking my way and I had to get home for Sunday lunch, but I did wait and I told him what I'd seen, which included Lesser Whitethroat and various other things. He said to me, 'You seem to know a lot about the Park. How would you like to be the official bird observer?' I said that would be interesting, but I'd no idea how to get the job, and he said it was easy because he'd just appoint me, as he was the chairman of the committee. The man was Lord Hurcomb, who had been President of the RSPB. He was about ninety, and he didn't realise I was only thirteen. I should have been eighteen to do the role, but he got me signed up and it cemented my interest in recording and purposeful birding – and probably stopped me becoming a twitcher, which was the obvious thing to do.

When I was at school I did get bullied a bit for being a birdwatcher, as it wasn't seen as being a cool thing to do, but I didn't let that deter me, and I set up a YOC group in Thames Valley Grammar School. None of the boys joined, but five girls did. We were all about fourteen, and the biology teacher, Mrs Boucher, used to take us out on walks. I would be the organiser and we'd very often go to Bushy Park or Richmond Park. Eventually the boys disliked me even more because I was spending my time with a load of girls they wanted to go and meet, but they weren't members of this group so they were cut out.

I started doing the Common Birds Census in 1976 when I was sixteen, and went to my first BTO conference the same year. It was clear to me that I was always going to be into birds and at one point I did think I'd like to be a reserve warden, so I got leaflets about how to do it. Someone said to me that I'd never earn enough money being a reserve warden to have the kind of life I wanted to have and enjoy birdwatching – I was told it was much better to get a job doing something else which paid better and to keep birdwatching as my hobby.

And that's more or less how it worked out. I left school at eighteen, wasn't quite bright enough to go to university, got a job at Shell International and persuaded them to fund me to do an HNC in Business Studies for two years, which I did for two nights a week to get the qualification. I started in marketing, thinking that it sounded good and was probably what I wanted to do, and eventually worked my way into public relations with Shell. I then left Shell to organise exhibitions for British Telecom, and then worked for a public relations company. I remember going for the interview with the PR company. They asked me what I knew about the Association of British Travel Agents (ABTA), which was one of their clients, and I said, 'Not much, really.' They asked me what I knew about travel and I said I loved travelling, I ran birdwatching trips to France at the weekends. I remember thinking that was pretty stupid as I was telling my potential employer that instead of working for them I would be spending time scheming to go on holiday. So I started to backtrack a bit, but they said it was fine because they wanted someone who knew about travelling, so they asked me to work for them and work as an intern at ABTA. Three years later, ABTA decided it was going to let go of the PR company but asked me if I'd join them and work directly for them. So I ended up working for ABTA for a further seventeen years. It was great because I ended up travelling the world as part of my job. Birdwatching was easier to do as a result of that, and I built up quite a nice bird list.

What is your life list?

It is heading for 7,600. I've just been to Cameroon and added eighty-two species, and the month before that I was in Brazil and added about ninety. I'm off to India in a month and will probably add another fifty there.

What would you like it to be?

Well, there are about 10,500 species if you use the International Ornithological Congress list – which I do – and the highest-scoring person at the moment is probably on about 9,100. I don't think I'll reach 9,000 because you need time, dedication, money and good health. Time – who knows whether I've got much of that? Dedication – I probably have got that. Money – well, maybe, I don't know. And health – again, I don't know. Currently I might spend £5,000 and go and see eighty-two new birds, and that feels good, but if I spent that amount and saw eight new birds I probably wouldn't feel good about it and I'd get a bigger kick out of putting £5,000 towards something to do with nature conservation. So I think I'll get to 8,000 in about three or four years, although I'd quite like to get to the same level as Phoebe Snetsinger got to, which is just over 8,300. I'll probably review things at 8,000 – see how much I'm enjoying it, see what's left. I'm nearly fifty-four now and don't particularly want to spend my sixties climbing mountains, sliding down slopes and sleeping in rubbish hotels – I'd rather concentrate on seeing some birds closer to home or places that are a bit more enjoyable to go to. Right now I'm trying to get through some of the horrible and unpleasant places.

I've seen far fewer species than you – probably about 1,500 – and I'm not sure I can remember all of them. Do you remember?

I remember about a third of them well. Of the remaining two-thirds, about half of those I can vaguely remember and would be able to put a name to given a bit of time, and the other half would be birds I can't actually remember at all, and probably wouldn't be able to name even if I had the book in front of me and they were there. An awful lot of the birds I've seen around the world have been shown to me by other people. A lot of my friends joke about it because I like wildfowl and out of the 180 or so species in the world I've only got twelve species left to see, which is probably more than Peter Scott ever got to see. There are some wonderful wildfowl left to find – such as the Brazilian Merganser. I've added White-winged Duck and Hartlaub's Duck to my list in the past year. The toughest one will be the Madagascar Pochard, but now that people have managed to bring it back from the brink of extinction it would be crazy not to see it.

Some people would think it slightly odd to pay so much attention – as a lot of birders do – to how many species they've seen in their lives, as a lot of the memories must be slightly hazy. Isn't it a bit like Don Juan sleeping with thousands of women and not being able to remember most of them?

Well, I think if you slept with over 7,500 people and you could remember 2,000 of them by name, that's not bad going! But you might wonder if the other 5,000 were really worth it. Well, maybe they weren't – maybe they were dull and uninteresting. It's like drinking wine – some you remember, some you don't, but it doesn't necessarily mean the wines were bad, just not memorable.

Which species would you most like to see before you die?

The one I want to see more than any other – and I'd be really disappointed, by the way, if someone told me I was going to die before I saw this bird – is the

Emperor Penguin. And that's because its lifestyle is fascinating: the way it can dive to amazing depths and can hold its breath for a huge length of time. It nests on the ice but it walks all the way there because it wants the ice to melt right up to the point where it nests, in time for its chicks to then just jump into the sea. It's a hard bird to see. It costs about £15,000 because you have to go to a place called Snow Hill Island in the Antarctic to see them. I will do it, but spending £15,000 on one birding trip is quite a lot. On the other hand, though, putting it off is a bit stupid because it'll then cost £17,000 and it'll be even harder – and I'll probably have less money by that time as well.

Do you ever think about the amount of carbon you're emitting into the atmosphere with all your travelling?

I've got a very big carbon footprint. I've probably flown between two and three million miles now, both in my professional and birding life, which is an awful lot more than most people, and if everybody did that, it would be bad. I've never worried too much about my carbon footprint, though, because in comparison with the bigger things that cause carbon to go into the atmosphere – such as power generation, heavy industry and so on – it's a bit like rearranging the deckchairs on the *Titanic* to worry about my carbon footprint and those of other people who go flying, when air travel probably contributes about 2–3% to the total. And if you did halve the amount of air travel, you probably wouldn't notice the 1% improvement, but you would notice the world economy collapse.

You used the phrase 'purposeful birding' earlier, which is what you've done a lot of, and you've been involved in various ways with the organisations that organise birders into helping count birds. You've been on the BTO Council and the RSPB Council, and at twenty-two you were the youngest ever President of the London Natural History Society. Have you enjoyed all that because you enjoy people, birding and organising things?

It's quite odd that I chose a career in public relations which means that you have to get on with people. I do find some people very frustrating, and I choose not to be with people more than I choose to be with them.

I remember when I was with Peter Holden in 1974, when he was the national organiser of the YOC. He came from south-west London originally, so he came back there and we went for a walk in Bushy Park, me and ten other kids. If you'd asked me at the time who I most respected and would want to be, I'd want to be Peter Holden. I said to him – and he reminded me of this the other day – that one day I'd be President of the RSPB. I don't know why I said that. It could have been because I'd met Lord Hurcomb, who'd been RSPB President, but there was obviously something in me which quite liked the idea of doing things in a purposeful way and being involved.

When I was fifteen I gave a talk to the local Women's Institute, which was organised by Trevor Gunton of the RSPB. He got good feedback on it and asked if I'd do two more – he'd no idea that I was actually fifteen years old. I went to the RSPB to meet him and that was when he discovered my age, but by that time I'd done at least six talks which had gone well so there was no problem about it. It was then decided there should be a local RSPB group in the Twickenham area and after an initial meeting in 1978 I was invited to go along to a discussion to hear about it further, and where they were going to try and get a committee together. At the meeting Rob Hume and Mike Chandler of the RSPB said they wanted to have a committee but also asked for a volunteer to be leader. Of course, nobody offered, so someone said, 'Well, Keith, why don't you do it?' and I said, 'Well, if you want,' and I could see Rob Hume and Mike Chandler looking at the floor trying to think of a way out of this, as clearly they didn't want an eighteen-year-old running an RSPB group. But everybody in the room supported the idea and said they'd help me.

So I ended up running the RSPB Richmond and Twickenham group, which was, and still is, a successful group. I do their anniversary lectures every five years and I'm giving the 35th anniversary one in December 2014. I enjoyed running the group – my parents had always been quite good at organising things, such as school fetes, so they helped me. I think the RSPB was a bit hesitant about giving me a float to get it started, so Dad lent me £100, I hired a church hall for three months, put the programme together, got the speakers organised – and we had over 200 people at the first meeting. It wasn't long before I paid back the £100 to Dad, and three years later we gave the RSPB a cheque for £3,000. That was a lot of money in those days!

The RSPB made films and I really wanted them to put on a show in Twickenham but the person in charge of this at their headquarters said that it wouldn't be successful and refused to do it. Well, if anyone tells me I can't do something, that just makes me even more determined to do it, so I hired the films directly and publicised the event. I hired a hall that would seat 350 people. The woman at the RSPB was furious, and said it would fail. She was wrong. Over 400 people turned up, and at least fifty of them had to stand – but nobody seemed to mind. Back in the 1980s, film shows were all the rage. I think today you'd only get forty people!

I retired from the job in 1985 aged twenty-five, and I think it was a very useful experience. I learned how to get on with people, how to persuade them to do things they didn't want to do, and I probably got more out of that in terms of organising teams of people than I got from any kind of training later in my career.

You got into all that at a very young age. Later in life, you've been on the councils of quite a few bodies.

I've been on the BTO Council twice, once when I was about thirty and again recently. The RSPB Chairman, William Wilkinson, nominated me for a place on the RSPB Council when I was twenty-five, but I was also being headhunted for a job working for the Nature Conservancy Council so I reckoned it wasn't a good thing to do both. Unfortunately the job offer fell through so I contacted the RSPB again, but it was too late.

Do you think you made a difference when you were on the BTO and RSPB councils? Would you recommend it to others?

I think people should do it if they can bring something to the table which isn't there already. I think I wouldn't have been that useful on the RSPB Council at twenty-five and I'm glad I didn't do it, as it might have put me off – and put people off me!

On the BTO Council the first time round I learned a lot, and I think I got more from it than it got from me. I joined the RSPB Council when I was forty-four – and again I learned a lot from my first two years, and if I gave anything back it would have been in my last three years because it takes a while to get your feet under the table and learn how it works. I hope I made the RSPB think about some of the things it was doing about its membership. When I had my second spell on the BTO Council, though, I think the BTO needed to change the way it talked. It was an organisation that could easily feel that it was a bunch of people wearing white coats talking about birds to birdwatchers in a way that scientists, rather than birdwatchers, spoke. So I tried to help them think about changing that, and about changing the logo to something less corporate and more 'fun', changing their attitude of looking at members as an important resource rather than just being there but not adding much value. There were some people in the BTO who did see that value, and others who didn't.

Graham Wynne, the RSPB Chief Executive, wrote to me when I left the RSPB Council, thanking me for my time and saying it had been 'very useful and challenging against a background of total support' which I thought was quite an interesting phrase. I suppose I can be a bit direct at times – which can be good and bad. At least everyone knows where they stand!

Have you been into bird ringing at all?

I went to Queen Mary Reservoir in Surrey in the summer school holidays when I was about fourteen and met a man there doing bird ringing. I asked him if he minded if I watched him. I think he said, 'Well, yes I do, actually.' He was quite unwelcoming and arrogant, so I didn't feel at all that this was something I wanted to do. I never got into ringing and probably never will. I haven't really got the time to offer that it takes now to get to the right level of qualification and training. But I am very lucky that I study three species that people want to get close to: Stone Curlew, Peregrines and Red Kites, and I've had the opportunity to take chicks out of the nest and hold them in my hand when working with ringers, and I've got a lot of joy out of working closely with the birds in that way.

Have you ever been a twitcher?

As teenagers, most of my birding friends started twitching – all piling into a car to chase a rarity in Norfolk. I never did that. I used to go loyally on RSPB or London Natural History Society field trips. The only time I ever got into twitching a bit was when I was about twenty-nine, when a friend in the pub described me as a 'low-lister'. At that point I probably had about 300 species, when you really needed 400 to be considered a 'proper' birder. I was a bit wounded by the comment – and in fact probably had the lowest list of all my mates in the pub – so I did go on a bit of a mission, with him and the others, to try and fill in the gaps. For instance, I'd never seen a Black-winged Pratincole – which I did get, together with an enormous speeding fine as I had to go to Cornwall, and I went twitching for about six or seven years. Every Friday night in the pub we'd check our pagers and decide where we were going to go at the weekend. We might drive to Northumberland on the Saturday, come back, find there was something in Wales, all go to bed and set off there the next day. So I got myself up to 400 species – my 400th was Blyth's Pipit.

But it's a bit like when you've never had sex, everyone says you haven't lived yet, then when you've had sex you think, 'Well, what was all that about? It was OK, but my life hasn't changed, except that I'm no longer a virgin.' So when I got to 400 species, all I knew was that I now wanted to get to 500. So I'm still a low-lister. I've only seen 416 species in Britain because after the 400th, which was in 1998, I just didn't feel the need to go and chase rarities.

I do quite like the idea of getting new birds on my county list as I'm the county recorder for Hampshire, so I should really be in the top twenty birders in the county. But emotionally I'm more interested in seeing birds around the world, and on the Surfbirds (www.surfbirds.com) list I'm usually ranked as about 35th, but there are loads of people who don't declare their lists so I'm actually probably in the top 200.

Which bird book has influenced you most?

The AA Reader's Digest book was the first one I had. It had a big influence on me and I can still describe virtually every picture in that book. Prior to that, the Ladybird books and the series *What to Look for in Spring/Summer/ Autumn/Winter*. I don't know where they went, probably to a jumble sale or something, and a few years ago I bought them all again, as I realised that I'd chucked out a lot of my early bird books. I was very influenced by books by Eric Simms because I knew him and I liked the way he wrote. His New Naturalist on *Woodland Birds* I particularly enjoyed reading, and *Birds of Town and Suburb* was important to me as I was a London birdwatcher. I had the Hamlyn field guide by Arthur Singer, and the Heinzel, Fitter and Parslow guide, and of course Peterson, Mountfort and Hollom, the 1974 version, which was the one I used more than any other.

I was really lucky, as I got to meet Roger Tory Peterson in 1982. I was at the International Council for Bird Preservation World Conference in Cambridge and Jeffrey Boswell asked me if there was anyone I'd like to meet. I said I'd like to meet Roger Tory Peterson, so I dashed out and bought a copy of the field guide from the nearest bookshop in Cambridge, came back and I was introduced to him. It was very interesting: he had a little group of admirers surrounding him and he stopped talking to them and started talking to me when I was introduced to him. He talked to me for fifteen minutes and I could see all these other people were really irritated, as they now couldn't talk to him. I read his biography after he died and realised that when he was nineteen he went to the American Ornithologists' Union Congress and found it hard as he was the youngest there – I wonder if he thought that about me.

Is there anyone in birdwatching history you would like to have met?

I consider myself incredibly lucky to have sat in on meetings back in the 1970s with so many 'giants' of ornithology – Max Nicholson and Stanley Cramp, for example. But the one person I would choose is Richard Meinertzhagen. He died when I was seven, but I would love to challenge him about the fact that much of his life appears to have been a complete lie.

Do you have any other hobbies?

I like wine and I love looking at wine books. I've got just about every book on wine that's been published in the past ten years! If I'd been into football I'd probably have collected programmes, and if I hadn't been a birdwatcher I'd probably have been a bus-spotter as I think buses are more interesting than trains: they're on roads, you can do it from the car, and you don't have to stand by railway lines. I am a bit obsessive: I have files on my computer with trip reports from places I've never been to and may never go to. So county recorder for Hampshire is probably a good job for me because every year I sift through the records of about 800 people, which amounts to something like 100,000 records.

Would you say that you are ambitious?

I would not, but I am very driven. That's probably because when I was about eighteen I was aware that I hadn't done terribly well at school and I just about got through my exams on every occasion because I hadn't been driven. So not having gone to university and starting work at Shell, I became driven. However, I am ambitious for the charities I am involved in. I like raising money – there is nothing better than someone coming up to you and saying, 'Would you find it helpful if I gave you £10,000 for the African Bird Club?' That doesn't happen every day, of course – but it's happened three times in the past three years!

What's the best birding place you've been to?

Kenya – I did a self-guided trip there in 1983 when I was twenty-three. I saw 250 species and thought I'd done well. But in 2004 I went back with a tour operator and saw over 750 species!

Where's the worst place you've been birding?

The Maldives. I saw four species in six days – but I was sitting by the pool all day!

What's the best bird you've ever seen?

That is such a hard question, because my favourite bird is the one I've just seen. I know that sounds corny, but it's true. So right now that would be Grey-necked Picathartes (in Cameroon).

What's your favourite bird family?

Wildfowl – and I get my leg pulled a lot about this one!

What's your most wanted bird in Britain that you've not seen?

Black Lark! Rather than twitch the bird in Anglesey, I chaired a meeting of the African Bird Club.

If you could see one more bird in the world that you've never seen before, what would you choose?

Emperor Penguin – because of its amazing breeding cycle.

If you could have an airline ticket to somewhere in the world you haven't been, where would you go?

Antarctica – and I am going in November 2015!

Imagine you're on a desert island and you could have some music: what would you choose?

Barry White: All-time Greatest Hits.

What's your favourite film?

Skyfall.

Favourite TV programme?

Spooks.

A bird book to take to the desert island?

The Handbook of the Birds of the World (preferably on a Kindle!).

And a non-bird book?

The Times Complete History of the World.

ROGER RIDDINGTON

Roger Riddington is the editor of British Birds *magazine. He was born in the 1960s.*

INTERVIEWED BY KEITH BETTON

How old were you when you started getting interested in birds?

I must have been about ten. Up till then, my mother had been trying very hard to get me interested in birds, and of course my natural response was to do the opposite; football and other things were more important at that stage. But I was off school for a few days (in January/February 1977), which was rare for me; I got bored and resorted to making lists of birds coming to the feeder in the garden. It went from there, and I remember the last day I was off school I was well enough to go with Dad on his milk round in Lincolnshire, so my list got a bit bigger. It grew slowly from there – the secondary school I went to wasn't the sort where you told people you were interested in birds, so I was a closet birdwatcher while I was there and I decided when I went to university I was going to 'come out' and it wasn't going to be a secret any more! I suppose that, in the early days, my birding simply developed as a result of taking the dog out round the woods and fields, seeing hedgerow and farmland birds.

Who were the influences on you as a young boy in your birding?

To a large extent I was self-taught. The one person I would credit with my birding development in my first years until I finished secondary school was Ted Smith, who is now President of the Lincolnshire Wildlife Trust and has had an amazing influence on wildlife conservation in Britain generally. He lived (and still lives) in the same village as my mum and dad, went to school with my dad, and used to take me along to Lincolnshire Wildlife Trust conservation working parties in the winter, mainly woodland work (coppice restoration and maintenance). I thought it was fantastic, as you could use a hedge knife with a 12-inch long blade and help the guys wielding the chainsaws. We used to build huge bonfires with brush, which was great fun, and he'd teach me about wildlife in general, but particularly birds.

Did you have binoculars when you were young?

I had a pair of Boots Pacer 8×30, from about Christmas 1977, then some Hilkinson 10×50 in the early 80s. My first decent pair were Zeiss Dialyt 10×40, which I got for my twenty-first birthday in 1987. I got through a lot when I was on Fair Isle: first the Dialyts, then some Bausch & Lomb 8×42, then Swarovski 10×42. I got my first pair of Leica 8×32 BA in 1995 and I got through three or four pairs of them. Shetland's pretty hard on binoculars as they get filled up with salt. I remember sending one pair back to Leica and they said they had them on the work bench and they literally split in two! In 2008 Zeiss very kindly gave me a pair of their 8×42 FL, which are still my birding bins, but the latest Leica 8×32 are still up there with them, and they go on trips with me because they are very compact.

What about telescopes?

My first one was a Bushnell Spacemaster, an eighteenth-birthday present in October 1984. I'd looked through my first telescope a month earlier at Gibraltar Point in Lincolnshire, where there was a Wilson's Phalarope on the back of the Mere, and someone let me look through a Kowa TS1; I was really impressed. The Spacemaster lasted me till I went to Spain in 1989 and somebody pinched it from my car. My current one is the big Swarovski.

Did you ever think of doing other things apart from birdwatching?

I suppose you could say that birds have been my job all my working life, really. I went to Oxford and read geography because that's what I enjoyed, but I had no real idea of what to do afterwards. In 1988, when I finished, it was a time when it was really easy to get jobs in the City so I went for interviews with big firms of accountants. I went for three interviews and got three job offers, and I was within three weeks of starting work with Arthur Andersen in Cambridge when funding came through for me to do a PhD on the back of my degree. That led me to ditch accountancy and carry on at Oxford, doing a PhD on Great Tits, looking at movement and dispersal (so it related to geography to some degree). Andy Gosler was my supervisor, and Chris Perrins was one of my examiners.

When my PhD was finished I thought about various things – I'd gone straight from school to my first degree, and from there to my second degree, so I felt like taking some time off and spending a summer doing something else. I wrote to every bird observatory in Britain, but only two responded: Fair Isle (where the job was for assistant warden) and Cape Clear (where the job was for warden), and I chose Fair Isle. In 1992 I went to Fair Isle for the season, from April to October, and I was part of a fantastic team – Paul Harvey was the warden and Steve Votier was my co-assistant warden.

I've never really had a career path – apart from briefly at Arthur Andersen! I did enjoy maths and I was reasonably good at it, and I think my mother always thought I should be doing a job where I wore a suit and tie and earned good money.

I had a gap in my Fair Isle career because at the end of my eight months as assistant warden I had an opportunity to do a post-doc in Norfolk. I took over from Juliet Vickery who took a lectureship in Edinburgh and I joined a team of researchers working on Brent Geese at the University of East Anglia (with fieldwork in north Norfolk). Academically, that was good; I published more papers on Brent Geese in one year than I did on Great Tits in three.

The warden's job on Fair Isle became vacant from the beginning of 1994, which coincided with the end of my contract, so I did four years as warden on Fair Isle. My then partner was Australian, and three years was enough for her, so she returned to Australia in early 1997. The plan was for me to do another year on Fair Isle and then go to Australia, but for a number of reasons that didn't work out, so we split up and I finished on Fair Isle at the end of 1997. In some ways it would have been nice to spend more time on Fair Isle. I was fit enough and keen enough, but I made the decision to leave as there were some interesting job opportunities coming up at that time, and I ended up managing the Biological Records Centre in Shetland for three years. It was a completely new post, set up with European money, and while it was a great contrast to the Fair Isle job, as it was primarily indoors, it was a chance to broaden my education in terms of natural history, as there was a lot of interesting plant work.

In the middle of June 2000 I saw an advert in *British Birds* and I remember thinking, 'I could do that in Shetland.' I sent in an application and Richard Chandler phoned me in September 2000 and said that the board would like to talk to me. A major snag was that I was going on a month-long trip to China ten days later, but he very kindly said they would talk to me when I got back. And that was my last big trip.

Is the job full-time for you?

Yes. If you don't want to take any holidays, it's a comfortable five-days-a-week job, but if you want to take time off you have to get ahead before you go and catch up when you get back.

Let's go back to when you were a teenager. So you birded alone most of the time?

Yes. I joined the YOC in the late 1970s.

What was your first bird book?

An Octopus book. There was a series of four. I think there was one on garden birds, one on woodland and field birds, one coast and sea and one marshland, and they were produced in Europe so they contained quite a few birds that I'd never heard of or had any hope of seeing, such as Ortolan Bunting and Purple Heron. The first 'real' book I remember was the Mitchell Beazley *Bird Watcher's Pocket Guide*, which I loved because it had several pictures of each species and the illustrations showed jizz pretty well, and the *Popular Handbook of British Birds* by Phil Hollom – the one covering common birds (which got covered in sticky-backed plastic and was well-thumbed) and the 'rarer' one a few years later.

Where did you bird locally?

As well as the fields and woods around home, the nearby coast and Gibraltar Point. In the earliest days the closest point on the coast, about seven miles from where I lived, was a little village called Chapel St Leonards, which most birders will know these days from having an Audouin's Gull three or four years ago. Huttoft was two or three miles further north.

Did you think of getting a job at Gibraltar Point?

I'm not sure there were any! Work was helping Dad on his round. He was self-employed so it brought in some pocket money. I was a keen cyclist so it was no sweat to cycle to the coast.

Did you ever get to Norfolk?

We had a family holiday to Norfolk in 1981, when I ticked off Avocet, Spotted Redshank and things like that, but my first real trip was August 1985, the summer I left school. I went with some friends who had a caravan near Holt and they were kind enough to stick my bike in the back of the caravan and I slept under the awning and cycled off to the coast. The first day I was there was amazing – there was a Pectoral Sandpiper and a White-rumped Sandpiper on the reserve, every hide was chock full and I spent all day trying to see those two waders properly. I don't remember which hide I was in, but I remember very clearly that at about 4.30 in the afternoon the hide door opened and some bloke stuck his head in and said there was a Little Whimbrel at Blakeney. The hide emptied just like that and I thought, 'Great; it will be easier to see these birds properly now' (which it was). About two hours later I caved in, cycled off to Blakeney and saw the Little Whimbrel. So it was an amazing first day of proper birding in Norfolk. Up till then it had all been Lincolnshire-based and I'd never really been anywhere else, so when I went to Oxford, things like Nuthatch and Tree Pipit were all new.

What about overseas trips?

My first overseas birding trip was in 1986 when I went to Iceland for a month on a geographical expedition, working on a rock glacier. The deal was that you did ten or twelve days of slave labour, measuring rocks on the glacier – we camped on the glacier – but then we had two weeks of our own time and they gave us round-the-island bus passes. There wasn't much species variety but there were lots of birds: lots of Red-necked Phalarope, Harlequin, Barrow's Goldeneye (and my first Puffin!).

How many countries have you been to, and what's your British list?

About fifteen countries? No idea about the British list. I kept one till I moved to Shetland and now I keep a Shetland list which is possibly pushing 385, so I guess my British list must be about 450. I have absolutely no idea what my world list is.

Did you keep a notebook when you were an early birder?

Yes, and I still have them.

Did you do any sketching?

Yes, but I was no good, and I do less of it now than I used to, because photography has progressed so much. I regret that in some ways, because you look much harder at a bird if you're trying to draw it. It's so easy to just look through the back of the DSLR and fire away without really looking at the bird instead of trying to get the exposure and so on right.

Do you ever do any sound recordings?

I'm planning to start soon!

Are there any birders from the past twenty or thirty years that you haven't met and would really like to?

Bobby Tulloch, Ken Williamson (the first warden of Fair Isle), and Peter Davies (the second).

Who influenced you most of all?

In terms of ID, Paul Harvey has been a real influence on my birding; he's an ace birder. And of the people you look up to, I know Killian Mullarney a little: I've never been birding with him but whenever I've seen him at the Birdfair we've talked about identification and stuff – he's someone I have huge respect for. Ian Wallace is a hero in a different kind of way; I just think what he's done for British birdwatching is amazing. Ian Newton is one of my heroes as well.

If you cast your mind forward ten years, do you envisage still being editor of *British Birds*?

It's probably not a good idea for me to stay in the job for another ten years. I think it will then need someone younger and more dynamic and more 'e-savvy'.

What do you think birding will look like in ten years' time?

I wonder if we really will be travelling less and thinking more about our carbon footprint – I have a sneaking suspicion we won't be. Technology will have moved on, but how far I don't really know. I don't think optics will move on significantly. I think the web will become ever more important: people who aren't utilising the web (and birders who aren't taking photos or sound recordings) will to some extent become sidelined. I think rarity assessment and record assessment will to a certain degree be based on hard evidence – by which I mean photos and sound recordings rather than descriptions. I have mixed feelings about that; I wonder if in a century people will look back at our generation and think, 'They were dodgy observers, because they had nothing to back them up between specimens and photos/sound recordings.'

On that note, have you ever strung a bird which has ended up in print and is still in print, not having been corrected?

I can't think of anything that hasn't been corrected quite quickly.

Best birding place you've ever been?

Fair Isle.

If I could give you a chance to go birding anywhere for a day, where would that be?

I'd go to the Spoon-billed Sandpiper breeding grounds on the Chukchi Peninsula in Russia.

If I could give you the chance to see any bird in the world you haven't seen, what would that be?

Spoon-billed Sandpiper!

Of the birds you have seen, which would be the 'best' one?

The last rarity.

Favourite music?

Malachy Tallack's current band.

Films?

I'm not a film fan, really.

TV?

Sport in general, probably football mostly – I'm a Manchester United fan and have been since before they were good.

If you had to choose two books to take to a desert island, one a bird book (not *HBW* or *BWP*) and one not, what would you choose?

Lars Jonsson's *Birds: Paintings from a Near Horizon*.

A non-bird book?

Something by Nigel Slater.

Can you remember your best ever day's birding?

Friday 14 September 2001 – the day we found a first-winter Thick-billed Warbler on Out Skerries, a day with a north-westerly wind. I had three really good birding mates with me – Pete Ellis, Paul Harvey and Ken Shaw. We set off with no real expectations and had been birding round the islands. About two hours in, the bird hopped up on a dyke. It was the third British record and the first since 1971!

And your worst day's birding?

Getting my bins and scope nicked in Seville at the end of my first day in Spain in February 1989 wasn't great!

IAN NEWTON

Professor Ian Newton OBE, FRS, FRSE is one of the foremost British ecologists and ornithologists. He was born in the 1940s.

INTERVIEWED BY MARK AVERY

Would you call yourself a birder?

It depends how you define a birder. I don't go out to watch birds systematically. But if we have birdy visitors, particularly from abroad, then I enjoy taking them around and showing them things. And if I am abroad myself, I take every opportunity to see birds that I'm not familiar with, so that's when I do most of my birding. But I am a trophy hunter at heart. When I was a little kid I used to collect birds' eggs, but then I switched to collecting data instead. And I do ring birds, as you know, and that means that I handle birds in the garden most days of my life.

Really? You catch birds most days?

Most days, yes – if I'm at home and if the weather is good. Songbirds, obviously.

Is that for fun or is it because you feel you are contributing to the wealth of knowledge?

Mainly for fun!

That's very honest of you – I think most ringers would say it was because of the contribution to science.

I take a lot of information from each bird I get. And I have written a number of papers in recent years on the birds I've caught in the garden. The papers are mainly on moult, because it's straightforward to record and little-studied.

Can you persuade me that moult is interesting?

I doubt that! But I find it fascinating. It's a major annual event in a bird's life about which we know very little. It's the timing of moult in the annual cycle – when they do it – that interests me. Birds usually breed at the best time of year from the food point of view, as you know, and then many species moult immediately after breeding. But some migrants wait until after migration and moult in their winter quarters. So if you look at bird species as a whole they show great variation in the timing and duration of moult, and also in the sequence in which they replace their feathers. Such information is generally recorded in places like *The Birds of the Western Palearctic*, but few have much interest in it and moult is generally regarded as being one of the dull aspects of bird biology.

That's where I am!

I think that's where most people are. We have lots of books, for example, about bird breeding and migration, but how many books can you think of about bird moult? I know of one, written by a Russian and published in 1966.

Are you thinking of writing another one?

It's too big a job, and it probably wouldn't sell.

I can't argue with that! So how long have you been interested in birds? Has it been a lifetime interest?

More or less a lifetime interest. I was engrossed initially, at the age of five or six – as far back as I can remember – in mammals, but then I switched to birds. The thing about mammals is that most of them are nocturnal and so you can't see them in the same way that you can see most birds. Birds are much more 'in your face'.

And I was interested in their eggs. I think this started when my dad showed me a Blackbird's nest with eggs. Of course, at the time when I was brought up, near Chesterfield in Derbyshire, most boys collected birds' eggs – it was legal then, of course, but its legal status was not particularly relevant. We all did it.

Well, that's another thing I haven't done much of …

Well, you're a different generation, you're younger! I was in the last generation of egg-collecting children. Before me, my father did it as a boy, and all the professional ornithologists I know who were older than me, like Derek Ratcliffe, all collected birds' eggs as children. I stopped when I was about thirteen, I suppose. But nonetheless in those years I learned a lot about bird biology – breeding seasons, where they nest, incubation periods – all that sort of thing, I knew in my early teens.

Do you think modern-day ornithologists are worse off without that first-hand experience?

It's difficult to judge. The world has changed enormously since I was a kid. Most of my life was spent, when not at school, outside in the countryside, on an exercise of self-discovery. The information wasn't readily available and you taught yourself. Home was a place where you went to be fed and to sleep! Apart from that you were outside. And when I compare that sort of life with that of my own grandchildren – they virtually never go outside unless they are taken by their parents, and everything they see and everything they do is under parental or at least adult guidance. That's very different from my time. Now, whether that's a good thing or a bad thing I wouldn't like to say. But certainly when I was starting a PhD, I felt that I knew a lot about the natural history of the birds on which I was working. It is different nowadays, and students have different skills. Some may be very good birdwatchers, but as far as natural history experience and knowledge of the biology of the birds go, they probably have less, but they make up for it in other ways, such as computing skills.

Did you have an egg collection of your own?

Yes, I did. Enough to fill a shoebox (or the bottom of a shoebox, lying on cotton wool). And then I graduated, as did only the very keen, to having two shoeboxes. I had all the generally available species, with the rarest ones being Sparrowhawk and Kestrel, I would think. I had Snipe, which I was very pleased to get, and Redshank. And I got some seabird eggs, which in those days were on sale in coastal places. I bought mine in Whitby, if I remember rightly, on a school holiday. They were gull and Guillemot eggs and these were for sale in butchers' shops for food. I was able to add some to my collection that way (hence the second shoebox, as they were big eggs).

The other experience I had in my teens was that my dad started to keep canaries. He'd done so when he was a boy, and he decided he'd like to keep them again, in the shed in the garden. He said I could keep finches in part of the shed. I thought finches would be much more interesting than canaries, so I bought some.

I kept most of the British finches except rare ones like Hawfinch and Crossbill. And this got me interested in these birds, particularly in what they ate. And that's what really started me watching birds carefully in the wild. I had a very good idea of all the plants whose seeds finches ate. I used to bring home different plants and watch how different species of finches opened the seed heads. Being tame, you got a really good idea of how the different species dealt with the different plants. That early interest eventually led me into a PhD into the feeding habits of finches.

So as a boy you were finding nests of all the common species, like Chaffinch, Goldfinch, Greenfinch …

Not Goldfinch, actually. Goldfinches were very rare in north Derbyshire at that time. You saw occasional birds in winter but they were not common breeders where I lived. But I had a particular interest in birds of prey and finches, oddly enough, and in 1956 I found about ten Sparrowhawk nests within cycling distance of our house. I watched them and recorded their progress, and then the following year I checked the same woods and found no new Sparrowhawk nests. They were all gone. The whole population went in that short time, and I later learned that this decline was widespread, and that it followed the introduction of aldrin and dieldrin in agriculture. So I witnessed at first hand the effects of organochlorine pesticides on birds of prey.

You actually noticed that?

I noticed they'd gone, but I didn't know why – that came later. But I did notice that a few eggs in the ten nests of 1956 were broken, and I assumed they'd been attacked by a Jay or other predator. It was only later that I realised that I had witnessed the effects of eggshell-thinning. I was only sixteen then.

From school I went to read zoology at Bristol University and in my first term there, a notice appeared about a conference – the EGI (Edward Grey Institute of Field Ornithology) conference to be held in Oxford in the new year. This was a conference which David Lack, the director of the EGI, organised every year for students. That was in January 1959 and the notice invited students to give twenty-minute talks on their own observations. I thought that I could give a talk about the food of Bullfinches. I had never heard a lecture in my life (except university lectures), and I'd never been to a conference, but in my talk I just worked through the year, starting in January, listing everything that Bullfinches ate, which foods were important and how they dealt with them and so forth. Anyway, David Lack saw me afterwards and asked me whether I'd like to consider writing it up. He said he would read it through for me before submission. So I went back to Bristol and wrote it up, in long hand.

And your writing is awful! Was it better then?

No! Anyway, I sent it to him, and he made a few suggestions, such as not just to write it as a single unit, but to have some sub-headings, such as Introduction,

Chris Packham in the New Forest, Hampshire, 1980

Chris Packham in Botswana, 2011

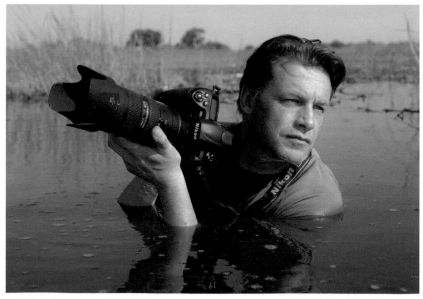

Phil Hollom with James Ferguson-Lees in Bulgaria, 1960 (Eric Hosking Charitable Trust)

Phil Hollom in France, 1974

Phil Hollom in 2009

Cover of *A Field Guide to the Birds of Britain and Europe*, 1974 edition

A Field Guide to the

BIRDS of BRITAIN and EUROPE

Roger Peterson, Guy Mountfort and P. A. D. Hollom

Precise field identification
of every species occurring in Europe
1225 Illustrations, 695 in colour
384 Distribution Maps, 384 Text pages

THIRD EDITION 1974

Stuart Winter in 1974

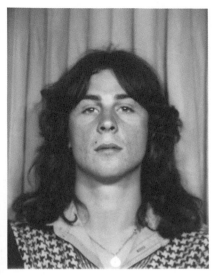

Stuart Winter in 2014

Lee Evans at the British Birdwatching Fair, 2009

Lee Evans holding a Woodcock, 2012

Steve Gantlett in Turkey, 1977

Steve Gantlett in 2004

Mark Cocker with local host in Morocco, 1979

Mark Cocker in 2014

Ian Wallace in 2012

Ian Wallace releasing a Leach's Petrel on St Kilda, 1956

Andy Clements in the BTO office, 2014

Andy Clements in 2014

Mike Clarke in 2014

Debbie Pain in Drakensberg, South Africa

Debbie Pain in Mexico, 2013

Keith Betton with juvenile Song Thrushes, 1975

Keith Betton in 1967

Keith Betton in Brazil, 2014 (Mark Tasker)

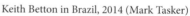

Roger Riddington in Shetland with Chris Orsman and Ann Prior, 1992

Roger Riddington climbing cliffs on Fair Isle
to ring Kittiwakes, early 1990s

Roger Riddington in 2014

Ian Newton with juvenile Peregrine, 1960s

Ian Newton in the 1980s

Ian Newton in 2014

Stephanie Tyler in Eritrea with her children, 1976

Stephanie Tyler with Bushbuck in Ethiopia, 1974

Stephanie Tyler at a camp in Botswana, 2010

Mark Avery with the over-14 WWT schools competition trophy in 1972/1973

Mark Avery and Chris Packham at Hen Harrier Day 2014 – soaked but happy! (Guy Shorrock)

Stephen Moss in the 1970s

Stephen Moss in the late 1960s feeding
House Sparrows, St James's Park, London

Stephen Moss feeding a Regent Bowerbird,
Australia, 2014

Stephen Moss in 2014

Ruth Miller in 2014

Alan Davies and Ruth Miller during *The Biggest Twitch*, 2008

Alan aged 19 with Peregrine, wardening at RSPB Haweswater, Cumbria

Alan Davies in 2014

Rebecca Nason in the 1970s

Rebecca Nason in 2014

Robert Gillmor in his studio, clockwise:
1973, 1977 and 1984

Robert Gillmor in 2014

Conclusion and that sort of thing – all those basic things that one should know. David Lack looked at it again, seemed to approve and suggested sending it to *Bird Study*. Mick Southern was the editor at the time, and he accepted it. I think it must have been published about 1960. I'm totally ashamed of that paper now. It was entirely descriptive – there was nothing quantitative in it – but the information was accurate and new. So that was my first EGI conference. After that one I went back in the next two years as an undergraduate and gave two further talks about other finches – Greenfinch and Linnet.

Were there lots of undergraduates giving talks at the EGI conference in those days? I admit I gave my first ever research talk at an EGI conference too.

I think a lot of people gave their first talks at EGI conferences. There were always some undergraduates, but mostly PhD students I think. I attended all of them, each year bar one (when I was in Canada) from then until I retired, and I presented a paper at almost all of them too. David Lack got the idea that I gave my first talk at that conference as a schoolboy, and for years afterwards, however many times I told him otherwise, he used to introduce me as the boy who first attended from school.

Although it's true that the work you described was done as a schoolboy while keeping and observing finches in the Derbyshire countryside.

Yes, that's true.

What do you think David Lack made of you?

I got to know David Lack through those conferences, and I asked him whether I could do a PhD on finches and he agreed. He must have been sufficiently impressed (or short on alternatives) to have me as a student. I was thrilled at the prospect and enjoyed my six years in Oxford, the last three as a post-doc.

How is your knowledge of plants, insects and everything else without wings?

Inadequate. I know quite a bit about plants – particularly those eaten by finches – and the common butterflies and so on, a bit about mammals, and a bit about fish because I used to go fishing, but not much beyond that.

Early influences?

I'd say my dad, for showing me that first Blackbird nest and for letting me keep finches. And after that, David Lack. I don't think there was anyone in between. I didn't know any other adults interested in birds. Obviously, most of the boys at primary school were interested in birds' eggs, but not in the creatures that laid them. And at grammar school I didn't know anyone really interested in birds.

First pair of binoculars?

I bought them out of my first student grant when I was nineteen. I had to wait until the end of the student year to make sure I would have enough money left over. They weren't brilliant. They cost about £20. But they kept me going into my thirties, and then I had a very good pair of Leitz Trinovids, if you remember those (8×40). Actually, I had two pairs in succession. And then I had a pair of Zeiss 10×40, which are what I have now. If I were to go for a new pair then I'd probably choose Swarovskis.

First bird book?

My first bird book was by Kirkman and Jourdain, called *British Birds*. I bought it when I was nine. It cost two guineas. I used to get a shilling a week pocket money and save it up. We were on holiday in Bridlington, and I saw that book in a bookshop and announced to my parents that I wanted to buy it. They discouraged me at the start of the holiday because they thought that blowing forty-two weeks of pocket money on a single book was not a good idea for what might be a very ephemeral interest. But I held on to the end of the week, looking at it in the shop window every day. At the end of the week I was allowed to go in and buy it as I hadn't in the interim spent any of my money. I loved that book. I must have read it any number of times – every species account. And it's got fantastic pictures of birds' eggs at the back.

Then I got T.A. Coward's book, *The Birds of the British Isles and Their Eggs*. There were a number of volumes but, conveniently, the birds of prey and the finches were in the same volume so that was the one I bought. And I got the Peterson, Mountfort and Hollom field guide as a school prize in 1957.

I see it says that you were in Lower Sixth Science 2.

When I was fourteen, I asked for *The Handbook of British Birds* as a Christmas present. It cost seven guineas then for the five volumes, and my parents got me the whole set. I spent the whole of that Christmas reading about birds of prey and finches, and being even more anti-social than usual.

Favourite bird? I expect you are going to tell me it's a finch, maybe a Goldfinch, or is it the Sparrowhawk?

It's got to be a bird that I've worked on: the more you get to know a bird, the more you like it. Either Bullfinch or Sparrowhawk, I think. Probably Sparrowhawk – it's more spectacular. But I do like Bullfinches too.

We'll come back to the Sparrowhawk, and probably the Bullfinch. One of your favourite species seems to eat some of your other favourite species. Favourite place to go birding?

We now live forty minutes from Rutland Water and I go there, mostly with visitors, several times a year. But I now also go to the RSPB reserve at Frampton Marsh which is only about twenty-five minutes away. It's very good; I always enjoy it there. This year I spent about half a day there with friends reading colour rings on Black-tailed Godwits. We passed them on to Jenny Gill and got all the details back, which was interesting. There's always something interesting to look at there. And the hides are very good – that circular hide is excellent. I think my favourite time there is late summer, when the godwits and Ruff are there.

Abroad, I think the place that I would probably go is Israel, where I've been many times, and particularly Eilat in migration time, because almost all the species which migrate between Europe and Africa can be seen there. The variety of birds is just incredible.

But the most exciting place for me is Arctic Canada – I'd love to go again. I spent a whole summer there in 1970 and that was the first time I'd seen Arctic waders displaying and nesting, and big colonies of Snow Geese and skuas. Snowy Owls – fantastic! I'd go back to Arctic Canada any time.

Even with the mosquitoes?

I'd try to go before they were active!

Do you have a birding/conservation hero?

That would have to be David Lack. His books and student conferences opened my eyes to bird ecology. He was my PhD supervisor but he also revolutionised so many aspects of bird ecology, evolution and behaviour. Every subject he touched he asked the right questions and pointed the way for so many of us. Clearly, he must be one of the top two or three most influential ornithologists of the twentieth century and so I count it as a great privilege to have known him. Lack was a bit remote to some people. He didn't easily relax. But I liked him and he was always helpful and encouraging to me.

His books are still heavily quoted. *Ecological Adaptations for Breeding in Birds* (published 1968) was really the start of behavioural ecology. It brought together every aspect of a bird's behaviour and ecology into a coherent whole that allowed you to understand things in a way that very few people had even thought about. That book lives on in the minds of many people. But the book that had the most influence on me was *The Natural Regulation of Animal Numbers* (1954), probably because I first read that at the right stage of life – as an undergraduate.

So there was a battle of ideas between Lack and Wynne-Edwards.

I lived through that.

And people of my age who came along after Lack had won the argument kind of got the impression that Lack was bright and right, and that Wynne-Edwards was wrong and clearly a bit stupid, but I bet it wasn't really like that.

No, it wasn't. Wynne-Edwards certainly wasn't stupid. He was Regius Professor of Zoology at Aberdeen University. As you know, the argument hinged on natural selection versus group selection as an explanation of animal behaviour. Wynne-Edwards's ideas depended on group selection, and no one could work out how it could possibly act. His idea was that certain populations would win out over others – populations that controlled their own numbers so as not to over-exploit their resources would replace ones that did not control their own numbers and periodically ate out their resources. But David Lack took the view that natural selection acts on individuals, not on populations, and any bird that behaves selfishly will probably benefit over those that restrain their reproduction. The consequence was that birds would occasionally eat out their food supplies, and that starvation would be frequent (as indeed is often observed in nature).

The argument lasted until David Lack published his 1966 book (*Population Studies of Birds*) which was four years after Wynne-Edwards's book (*Animal Dispersion in Relation to Social Behaviour*) in 1962. But David Lack was not the only one responsible for demolishing Wynne-Edwards's idea of group selection; there was also John Maynard-Smith at Sussex and other mathematical biologists. Genetically, it looked as though group selection just couldn't work, and increasingly it was forgotten from the late 1960s onwards.

You must have met Wynne-Edwards?

Yes, and he was another rather remote man, but my dealings with him were very pleasant. He came to an EGI conference on that conflict between him and

Lack and of course all of us students were there waiting for this clash of the giants. But it never actually happened, because Wynne-Edwards would not engage publicly in argument. He gave his talk explaining his ideas, and when challenged by anyone he would simply say, 'It's all in my book. Read the book and you will understand.' Of course, we'd all read the book but didn't understand it or believe it. He was not the sort of person who would have a stand-up row, and nor was Lack, so it was conducted in a fairly gentlemanly way.

Wynne-Edwards brought to Oxford with him people like David Jenkins and Adam Watson, who also talked at the conference. These were people whose work Wynne-Edwards interpreted as supporting his theory, but they didn't interpret their own work in the same way, so that was quite revealing.

Let's talk about Sparrowhawks, then – you've been interested in them since you were a child in Derbyshire.

Yes, but the opportunity to work on them really didn't come until Derek Ratcliffe had cracked the shell-thinning problem – and it came for me in 1971. In most of Britain there were very few Sparrowhawks left at that time and so it was difficult to find enough to study. In south-west Scotland there was still a reasonable population.

I thought that I would monitor numbers and nesting success to see how they responded to declining use of organochlorine pesticides. In my first year I thought it would be great if I could get my hands on these birds to ring them. So I made a box trap with a lower compartment containing one or two Sparrows to act as decoys and an upper compartment to catch the Sparrowhawk if it entered. Working as I did for the Nature Conservancy, I was allowed a licence to try this out.

I set up my trap on the ground near a nest, left it for about an hour and then returned to find a hawk inside! I can still feel the excitement. For then I knew I could catch them and I had the makings of a good population study. I'd already found about fifty or sixty active nests so that first capture changed things totally. I made more traps and caught about twenty Sparrowhawks in that first year as well as ringing the chicks. After that, I tried to catch the entire population each year. I never managed it, but I did get most of the females every year, enabling me, over a period of years, to follow the life histories of individuals. After a few years, I was joined in this work by Mick Marquiss, and this enabled us to extend the size of the study area to get larger samples, pooling our efforts and ideas.

What were your major findings?

The areas I worked in were well wooded. I found that the nests were regularly spaced, which suggested some sort of territorial pattern – which nobody had suggested before for Sparrowhawks. The interesting thing was that the spacing differed between areas. In fertile areas of the country, nests could be as close as half a mile apart, but in other areas the sites were up to several miles apart, but equally regularly spaced, so the question was, why the density differences between regions? I found that they were related to soil fertility and to the densities of songbird prey species. It turned out that Sparrowhawks were simply spacing their nests more widely in areas where woodland songbirds, their main prey, were at lowest densities. So this showed not only that Sparrowhawks were territorial, but that their densities were related to their food supplies. That seemed like a step forward, because I hadn't seen that type of data for any species before – so that was exciting.

From ringing the birds, we were able to follow individuals year after year through their lives. This allowed me to follow the annual survival and breeding success of birds at different ages. It was known that young birds were less successful than adult birds, but I don't think anyone had looked at individual performance on a year-by-year basis through life. And it showed that Sparrowhawks got progressively better at breeding and surviving up to about five or six years of age, and then as they got older their annual survival and breeding gradually deteriorated – indicating senescence. That was the first time that senescence had been demonstrated in a wild bird. The problem had been that birds can be aged only in their first year or two of life, but birds die at all ages throughout their lives, so you need a huge sample of juveniles to have enough old individuals to get a statistically significant result. Apart from Sparrowhawks, few bird species had been studied over long periods in which individuals had been ringed – Great Tits, Kittiwakes and Fulmars in Britain were notable exceptions – but no one had thought to look in detail for evidence of senescence.

I also managed to record the lifetime reproductive success of about 200 individual females – and that was another first. The interesting thing was that only a small proportion of the birds in any cohort produced most of the young in the next generation. This was mainly because most individuals didn't live very long, although it wasn't always the most long-lived birds that produced most young. The best-performing female Sparrowhawk I had lived to be eight years old but produced twenty-five offspring – some Sparrowhawks lived longer (up to twelve years) but produced fewer young. And about 16% of females that got as far as laying eggs produced no young at all in their lives.

You obviously like Sparrowhawks?

Yes!

And I like Sparrowhawks too, because I grew up at the time when they were rare. Even around Bristol where I lived, in the early 1970s, seeing a Sparrowhawk was quite unusual because of pesticides. I'd guess that I used to see them once every four or five times I went birdwatching, whereas now I'd see them almost every time. So a Sparrowhawk always, still, seems special to me, but not everyone feels like that about them.

People who dislike Sparrowhawks seem to fall into two categories. I like to take nature as it is without making value judgements. But some people dislike Sparrowhawks on emotional grounds. They may see a Sparrowhawk killing a songbird and eating it on their lawn and they don't like that. It seems cruel to them. They are reacting in an emotive way to nature. People like that are harmless, and I don't mind them, because they are not the people who kill Sparrowhawks and other raptors.

But don't you think there is something wrong with our education system when people see something as amazing as a predatory hawk and dislike it? And they don't seem to feel like that when they see a Blackbird pulling a worm out of the lawn.

Or a Barn Owl killing a rat, or a Puffin with a bill-full of little Sand Eels. Seeing one bird kill another does arouse people's emotions but that's how it is, and I don't mind that.

But then there's the other category: those who see birds of prey conflicting with their own interests – game shooting, in particular. And they are the ones who are damaging bird of prey populations, limiting the numbers of several species, and have done so for 150 years. And that I'm not so keen on.

You would say, would you, that game interests have limited, and are limiting, the numbers and distribution of some of our birds of prey?

Yes, undoubtedly. For some of our birds of prey I think there is no question of that.

Well, they say that they aren't ...

Well, that's strange, isn't it, when you can wander over a game-shooting estate and find traps and poison baits, and sometimes dead raptors. And nesting pairs disappear during the course of a breeding season in a way that just doesn't happen elsewhere.

I think it's a mixture of those two types of people who end up worrying that, because of Sparrowhawks, we end up seeing fewer songbirds in the countryside. Should we be worried at all that Sparrowhawks are affecting the numbers of farmland birds?

I actually don't know of any scientific evidence that Sparrowhawks lower the breeding densities of their songbird prey. I've looked at this myself and couldn't find any evidence. But, more importantly, the BTO looked at it, several times, with longer-term and larger-scale data than I had, and they couldn't find any evidence either. So I certainly don't think that Sparrowhawks are responsible for the recent declines in numbers of farmland birds – there's no question about that.

But even if it turned out that Sparrowhawks result in some species' populations being slightly lower than they otherwise would be, then I wouldn't be completely surprised at that – but the evidence doesn't show it, at least at the moment.

There's no question that Sparrowhawks eat a lot of songbirds – that's their primary diet and has been for millions of years. When you think of the reproductive rate of songbirds, take a Blue Tit, for example, each pair might raise ten young in a year. So, for every Blue Tit pair in April you could have twelve birds at the end of the breeding season. So if numbers are to remain stable in the long term, ten out of every twelve Blue Tits have to die of something each year – if not from predation, then from food shortage, disease or something else. So a lot of birds can be killed by Sparrowhawks without it impinging on their breeding numbers.

I don't remember us being up to our knees in Blue Tits while Sparrowhawks were absent.

Or other species too. Interestingly enough, Blue Tits and Great Tits have increased over the past fifty years, covering periods when Sparrowhawks were absent and when they had become abundant again, despite those species forming a significant part of the Sparrowhawk diet.

If the decline of farmland birds like Skylarks is not the fault of Sparrowhawks, then why have they declined?

Because their habitats and food supplies on farmland have been destroyed by changes in agricultural procedures. Anyone who was around in the 1960s and 1970s will remember the degree of hedgerow removal at that time. Think of all the birds that nested and fed in those hedgerows. I can also remember well when wet grassy fields were being drained for agriculture – those were great habitats for Lapwing, Snipe and Curlew but they barely exist now. Then there's the reduction in food for birds thanks to pesticide use. Herbicides remove what farmers call weeds, but these were the host plants for lots of insects, and also produced the seeds that many seed-eating birds depended upon. And fungicides remove the food supplies of fungal-feeding insects and some also kill insects directly; and then there are insecticides that kill a wide range of insects. So the insect fauna of farmland is now a small percentage of what it used to be, say in the 1950s and 1960s.

Do you ever go back to those parts of Derbyshire where you grew up, and what do you see?

I went back a few years ago and I saw far fewer songbirds than I remember being there in the past. I also saw more waterbirds, because there is more water there and the local river (the Rother) has been cleaned and now supports fish and herons. And more birds of prey – I saw Peregrine and Buzzard which I never used to see. But the once-common farmland birds were all pretty scarce.

Of course, the other evidence we have that things could be better for farmland wildlife comes from those places where the changes have been reversed and wildlife has come back again – places like the RSPB's Hope Farm, but there are others. I saw the Arundel Estate of the Duke of Norfolk in Sussex recently where Dick Potts has been working, and that is really rich in farmland birds, not just partridges, and there are raptors there too. It goes to show that if you have the food and the cover everything benefits – including the raptors.

Let's talk about another bird of prey – the Hen Harrier. You chaired the Langholm study, which investigated the impacts of Hen Harrier predation on the numbers of Red Grouse that were available for sport shooting. What did it show?

It showed me what a high density of Hen Harriers you can get if they aren't shot all the time. I think we started with two pairs of Hen Harriers and then, in the absence of persecution, they had increased ten-fold by five or six years later.

And that surprised all of us, I think ...

Well, it was good habitat for harriers, but it indicated the impact of illegal persecution on these birds.

You don't have any doubt about that?

The densities of Hen Harriers at Langholm were as good as anywhere at the start of the study, but when they were protected then the numbers increased.

So there we have an area where they weren't shot and they were breeding successfully and they reached very high densities quite quickly – and presumably that would happen in lots of other places if the harriers were protected. And there's lots of independent evidence that gamekeepers kill harriers on some grouse moors across the UK.

The second major lesson was the impact of sheep on these moors. Intensive sheep grazing really can destroy heather – and half the heather on that moor had gone in forty years or so (as shown by aerial photographs). If that trend had continued that moor would eventually have lost all its heather. Loss of heather could have accounted for the long-term decline in grouse bags at Langholm.

The third lesson was, given that sort of habitat, it looked as though it would be impossible to have driven grouse shooting in the presence of that number of harriers. That seemed scientifically convincing to me.

The mixture of grass and heather at Langholm attracted the harriers because it was so good for voles and pipits, and that's what the male harriers mostly eat in spring. If you go on moors where there is a lot of heather then you don't see many harriers – even if they aren't shot – so that harrier problem was partly the result of past land-use practices. A lot of moorland in Britain is in that degraded state now, so clearly we can't have driven grouse shooting and harriers at those densities.

Do you have any doubt that it is gamekeepers limiting Hen Harrier numbers and their distribution?

In so far as you can draw any conclusions from circumstantial evidence, then that's the conclusion I would draw. The persecution is at such a level that it holds densities at a much lower level than they would otherwise be.

Although it is illegal, killing Hen Harriers is a perfectly rational reaction by gamekeepers to the conflict between Hen Harriers and Red Grouse shooting. If you leave Hen Harriers alone they will eat a lot of the shootable surplus of Red Grouse, so grouse interests often don't leave them alone, they kill them. It's an intractable problem. Do you have any solutions to it?

Solutions have been proposed. You could say you would manage the habitat and get the heather back so that there were fewer voles and pipits and, as a result, fewer harriers. But that takes time, and how does the landowner make money in the interim? So that's a very long-term solution – a couple of decades or so.

Feeding Hen Harrier broods seems to work – it reduces the numbers of young Red Grouse that harriers kill, but I can't imagine that method being taken up on a national scale by gamekeepers. Really, I think you are left having to accept a third proposal: that harrier densities could be limited on grouse moors, to levels that allow some Hen Harriers to survive but allow driven grouse shooting to survive too. The idea was to move the eggs or chicks of some harriers, but the difficulty was in finding landowners willing to accept them. But I think that would be a potential solution. If we had one or two pairs on all grouse moors we would be better off, and would probably not need to translocate any if moor managers were allowed to destroy any extra nests above their quota. Then again, it could lead to suggestions of extending the method to other species. Sadly, there is mistrust on both sides – how many grouse moor managers would you trust on this after what's happened in the past?

At the moment, I think there is little enthusiasm for such a solution among grouse moor managers. They already have their own solution and most don't have any harriers.

Of course, there is another thing they could do, which is to allow more eagles to survive. There's no doubt that eagles prey on harriers sometimes, and also that harriers will avoid settling near eagles. Eagles eat grouse too, but having one pair of eagles might be better for the grouse moor manager than having several pairs of Hen Harriers. And the eagles would kill a few crows and young Foxes too, so there could be other benefits.

Just a last word on gamekeepers. Is it just the occasional bad apple who is involved in bird of prey persecution? Some people – some of my ex-colleagues at the RSPB, for example – would say that they are all at it.

My experience is almost entirely in Scotland and I would agree with your former colleagues at the RSPB and say that it is many gamekeepers – maybe the ones in the uplands are worse than those in the lowlands, but many gamekeepers will be killing birds of prey habitually.

Of course, even if they were willing to do so, many keepers wouldn't have to kill harriers every year simply because there are now so few of them.

Yes, indeed – there are no longer enough to occupy the available habitat. As far as harriers are concerned, I think the methods of killing have changed. In the 1990s, when the first Langholm study was done, most of the killing of harriers was done at the breeding places but now, you hear from a variety of sources, much of the killing occurs at winter roosts. A few gamekeepers get together and go to a winter roost and try to get as many as possible of the harriers as they come in during the evening. You can see how that, perhaps at a roost in the South Pennines, you could account for maybe a dozen birds in an evening and no one would be any the wiser. I think that a lot of that is going on. A few keepers are killing a lot of birds.

Two retired keepers told me what they were killing when they were working, and it was an awful lot of raptors – including Hen Harriers, Buzzards and Sparrowhawks.

Is the study of birds alive and well?

Yes, I think so, and it is pleasing that we have lots of good young ornithologists. The situation has changed during my lifetime. When I started, the Natural Environment Research Council was funding most ornithological work in Britain and now they are funding very little. That source of support has gone right down. Most of that sort of work is now done by the RSPB or the BTO, which get their funding from various sources. That's where the growth has been. And I think that will continue into the future.

In fact, the BTO is in a good position, because it has masses of long-term and spatial data which will enable all sorts of questions to be answered that you and I would be interested in, concerning what has changed and why.

Most people know you from your books – how many books have you written?

I've written eight and edited four more. The first was the New Naturalist *Finches*.

I bought that one ...

Good! I'm honoured.

... I think, when I was at school. I had to save up my pocket money for that. When did it come out?

1972.

Yes, I was fourteen, so it was from my pocket money.

A good investment! That was the first.

Which gave you the most pleasure?

I think the one that had the most impact was *Population Ecology of Raptors* because, by chance, it came out at a time when there was a lot of interest in raptors across the world. But the one that gave most pleasure in the writing is the one that is the least known – and has sold the fewest copies. And that's the one on bird biogeography (*The Speciation and Biogeography of Birds*, 2003) which is, to me, a very interesting area. What influences bird species distributions, and why does species composition change in the way it does across the globe? These are quite big questions, and trying to answer them took me further outside my home ground than any other book. You can explain the present only in terms of the past, so you have to understand something of earth history, the movements of the continents and, how islands formed and disappeared, and so on. Getting all that information together and delving into glacial cycles and palaeontology was really interesting. And I don't know of anyone else who has pulled those fields together – maybe Darwin and Wallace. I like to think the book has sold less well than my other books, not because it is a bad book, but because biogeography is not a fashionable subject.

Well, there are lots of interesting things in that book for birders, such as the fact that there are lots of wrens in the USA and just one wren, which is closely related to the American Winter Wren, occurs in Europe too. And when you go to Spain and Portugal it's nice to see Azure-winged Magpies and then one day you realise that there are Azure-winged Magpies in China too – but there are none anywhere in between. I had a marvellous trip a couple of years ago across the USA where I went from east coast to west coast. Halfway across, I realised that I had stopped seeing some species that I'd seen every day and I was starting to see some new species instead. One example was the Chimney Swift, which I saw every day from the first evening, when I was supping beer in Washington, DC right across to somewhere near Deadwood, South Dakota, and I never saw another Chimney Swift as I headed west after that. All the answers to those puzzles are in your book.

I've just finished another book – a New Naturalist called *Bird Populations* which deals with the factors that regulate bird numbers, such as food supply, disease, predation, nesting sites, etc.

You must find writing reasonably easy …

No, I find it difficult.

I'm surprised, because I think half the battle when writing about ideas and facts is to get it all straight in your head – and you, I would say, have these things straighter in your head than anyone else …

I agree you have to have them straight in your head, and I don't know how difficult other people find writing, but my aim is to achieve accuracy, crystal clarity and succinctness, at the same time as an interesting read, and that is not easy. Everything I write goes through more than a dozen drafts.

What time of day do you write?

Morning. Mostly morning, and a bit in the evening. I'm no good after lunch, so I go and do something else, in summer often in the garden.

Does it get in the way of ringing?

Ringing gets in the way of writing! The most straightforward book I wrote was *The Sparrowhawk* because I'd analysed most of the data and published most of it as papers so it meant organising things differently, but it was all my own work, whereas the most recent books I have written have involved a huge literature and reading lots of other people's work and that's much more difficult. And you have to condense a mass of information into a readable form.

I usually start by writing a draft of a chapter out of my head – and it might be very short if I don't know very much about that area. And then I start reading the literature and adding things to the chapter. I might end up with something three or four times too long, and then the really agonising work starts on reducing it down to a coherent whole.

Writing can be a pretty lonely job, holed up in an office for hours and years on your own.

Then why do you do it? You don't have to – you are retired and nobody makes much money out of producing books, so why bother?

There is the satisfaction of having done it, although that doesn't last very long. But there is also a measure of arrogance in it – there are aspects of our science that I feel I understand and can explain to people as well as anybody. Whether this is true or not, it gives me some motivation to continue.

No, I think that is fully justified, and I'm glad you said it otherwise I would have done.

In essence, the books I write are about the things that I would have loved to have known when I started. Now that we know some of the answers about bird populations and distributions, it's good to give something back and write it all down. The disheartening part is that so many young people now seem not to read books of this type, but rely instead on getting their information piecemeal (and without quality control) via the internet.

Do you have any advice for young people?

I hate giving advice to anyone. I would suggest trying hard to go to university because you probably won't make much of a difference unless you do. And after that, it's a question of ability, motivation and hard work: if you have those characteristics, you'll make a mark. But I've been lucky: I did a post-doc and then I got a job straight away, and that rarely happens now. In that year only two jobs came up, and I got one and Dick Potts got the other. I often think that I could have been in Dick's shoes and he in mine.

I was lucky to get a job, and lucky in my first boss and in my colleagues. And I've also been fortunate to have a supportive wife and family. All these things make a difference.

Do you have a favourite film?

Arthur (Dudley Moore).

A favourite book?

Under Milk Wood (Dylan Thomas).

Favourite TV programme?

Life on Earth (David Attenborough) (like me, not very contemporary).

And how about your favourite music?

Ravel's 'Boléro'.

STEPHANIE TYLER

Dr Stephanie Tyler – more familiar to birders as Steph – is
probably best known for her studies of Grey Wagtails and Dippers.
She was born in the 1940s.

INTERVIEWED BY KEITH BETTON

Where were you born?

I was born in Yorkshire, but Lincolnshire holds my earliest memories. We moved there after the war and I grew up in Woodhall Spa which was a brilliant spot: it has woodland, it's on the edge of heathland, near the Lincolnshire Wolds, and we used to spend lots of time at Skegness and Gibraltar Point in my early childhood.

And how old were you when you realised that birds were interesting?

I can't really remember a time when I wasn't interested in natural history, not just birds – anything that grew, or called, or flew, even when I was three or four years old. My parents weren't particularly interested. My dad had Henry Williamson and Richard Jefferies's books, and they were great because they encouraged me. They bought me books, but not a pair of binoculars. Because my parents were teachers, the idea of getting expensive items like binoculars didn't occur to them. I didn't get those till I went to university!

So you had to find birds without binoculars?

Yes, but I managed. It's amazing what you can do because you've got good eyesight at that age.

Do you remember anyone saying to you, 'Get involved in birdwatching'? It's just something you grew up with?

Yes, I loved the countryside, and kept a nature diary from as soon as I learned to read and write, I think.

Were either of your parents into birdwatching?

My dad was a historian and Latin teacher – in fact, both my parents were interested in the countryside, but not the details. I was just an avid plant person, a butterfly person, bird person, anything really. I joined the Lincolnshire Naturalists' Union, as it was then, and the Lincolnshire Trust for Nature Conservation – that must have been in my early teens.

In the days before you had binoculars, who identified the birds for you and told you what you were looking it?

No one. I had books and I'm self-taught, I guess.

What was your main book then?

I started off with *The Observer's Book of Birds*, like everybody else.

Were you particularly academic?

My parents both taught at a boys' prep school, where I was the only girl, which was pretty horrible, so I think my interest in nature was a way of escape.

Which was more important to you, plants or birds?

I think it was plants, but I was into everything. If I went for a walk I was looking at everything, but got to know plants very well. I had two wonderful black-and-white Penguin books on plants of the British Isles and I coloured each one in as I found it. I was a bit of a nerd.

Woodhall Spa was amazingly quiet and safe, and I could go off for walks on my own very easily for a day and nobody would worry.

What about university?

I went to Cambridge to study botany, zoology and geology. I was very narrow-minded, and that was all I wanted to do. I joined Cambridge Bird Club as soon as I got there and met people like Jeremy Brock, Daphne Watson and Geoff Gibbs, who took me to a shop to buy some binoculars. A group of us went most Sundays in a minibus to places like the Wash and other sites on the Norfolk coast.

Bill Thorpe and Robert Hinde invited me to do a PhD on Mynah Birds in India but I didn't want to go as I was about to get married, so Brian Bertram went instead and they came up with more funding for a PhD on the social behaviour and ecology of New Forest ponies, which turned out to be superb because I spent three years in the Forest wandering around the woods and heaths. That's when I became interested in Grey Wagtails because I was doing a lot of ringing. I used to ring at Lyndhurst sewage works and was catching quite a lot of Grey Wagtails. I had a flat in Lyndhurst and every weekend I drove up to Cambridge – I bought an old Ford banger for £80 and drove five hours to Cambridge to see Lindsay. He was studying to be a vet and did a lot of practice around Ringwood and Southampton in the university holidays, so we had lots of time in the Forest.

Would you preferred to have done a PhD on Grey Wagtails?

I should have done, shouldn't I! But there was no funding for that.

What was your main finding on New Forest ponies?

People hadn't really studied the social behaviour aspect. It was a dream subject because I could recognise individual horses and I used to follow foals from birth to two or three years. It sounds pretty boring, but the herd structure was interesting – it's a matriarchal group so there was a dominant female and various subordinate females. The dominant female has dominant offspring, and males and females behave very differently from an early age. I was very lucky because my thesis was published in 1972 in one of the *Animal Behaviour Monographs*.

Did you get involved with any of the Hampshire birdwatchers in those days?

I only got a little bit involved. I did a Common Birds Census in the New Forest but was flat out doing my PhD and driving back to Cambridge at weekends. Colin Tubbs mentored me to some extent.

Lindsay got a job in Chippenham, Wiltshire, so we moved there to an isolated cottage on a stud farm with Nightingales all around us and while we were there we had two children, Robert and Sally. I had a few temporary jobs with the Nature Conservancy Council doing woodland surveys and road verge surveys, but it was all contract work; I didn't have a full-time job. I was very involved with the local trust and, with Phil Horton, established a conservation group doing management on reserves.

After three years in Wiltshire we went to the United Arab Emirates for six months on an agricultural research station at Digdaga in Ras Al Khaimah State. The research station had lots of areas of Lucerne growing and irrigated channels at the foothills of the Oman Mountains. I did some ringing, and we just spent our time exploring and birding.

Michael Gallagher was stationed at Sharjah, about an hour up the coast near Dubai, and if we found anything exciting he would come down to see it. It was

Mike who persuaded me to join the British Ornithologists' Union (BOU) and British Ornithologists' Club (BOC). We also corresponded with Effie Warr. In 1973 we'd hoped to go to Tanzania but instead an opportunity came up for Lindsay to work in Ethiopia, and that's where I got to know John Ash very well. We were stationed in Addis Ababa for two years and spent a lot of time with John doing ringing. I was also able to study the wintering Grey Wagtails and local Mountain Wagtails. Then we moved up to Mek'ele in Tigray Province for our third year, where we ran into the revolution and other problems and unfortunately ended up staying an extra eight months because of that!

In 1976 the British ambassador came up to see us to see if it was safe enough to keep on working and he sanctioned us carrying on. Lindsay's job was looking after vaccinating teams trying to get rid of Rinderpest (an infectious disease of cattle), and we moved around from place to place. We went into the Danakil Depression, and climbed up Mount Ert 'Ale, an active volcano. Shortly after the ambassador's visit we were trying to get home after one trip and the Tigrayan People's Liberation Front had blown up the bridge so we couldn't go back via the main route. We decided to take a detour through the mountains and ran into an ambush. We were in a green Land Rover, so they thought we were the police or military and they fired on us. I remember being incensed, leaping out of the Land Rover and shouting, 'Little children!' in the local language. They kept us for a few days, said they were going to put us in a vehicle to supposedly release us, then when we were out of the village ordered us out at gunpoint and for the next month we just moved from place to place with them, gradually moving north to Eritrea. We spent seven months somewhere in the Red Sea Hills – we never knew where we were, as we always travelled at night.

Why did they keep you and not just let you go?

I don't know. We got off to a bad start because our vehicle had jerry cans on the top with diesel in – our vehicle was petrol and the diesel was for the vaccinating teams. They commandeered our vehicle and when it ran out of petrol they put the diesel in. They said they just wanted publicity for their cause.

There were four of you. Obviously they fed you, but did you become ill or lose weight?

We got dysentery from time to time, but they fed us adequately on unleavened bread and tins of peas and broad beans from China, but they did get milk powder for the children. It wasn't a very exciting diet! Sally was five and Rob was seven.

How much of it do the children remember today?

Sally doesn't really remember it at all, but it was quite traumatic for us. There were a lot of aircraft going overhead looking for targets, and we were shot at in a house we were sheltering in.

During the eight months did you ever fear for your lives?

There were a few occasions at the beginning, but after that we just believed we were of some value to them. We were captured in May and it was September when they captured John Swain, a reporter with *The Sunday Times*. They brought him to the same guerrilla camp we were in, and he was allowed an audience with them and he then went back to the UK and publicised our plight. We heard over the radio they were asking a ransom of $1 million but we knew the British government didn't pay out, and rightly so.

Were there times when you felt you couldn't carry on?

Early on it was difficult adapting to the conditions. It was very hot, the water was very salty and we were quite restricted in what we could do, but we were in a fabulous place. It would have been different had we been there of our own free will.

How did it end?

I think there had been lots of negotiations behind the scenes, with pressure being put on Sudan, from where the guerrillas got all their supplies, and eventually we were just told in January 1977 that we were going. They loaded us into a vehicle, we drove all night and slept on the sand, we then realised we were over the border into Sudan and not at risk. They took us to Port Sudan, bought us some clothes and got us a flight to Khartoum, where we spent the next three days in the Embassy until they could get us a flight back to the UK.

Recently John Swain phoned us because one of the guerrilla leaders has been living in Holland for some time, and he went over to see him. The ex-leader actually phoned us and we asked him why he had kept us. He apologised and said that he had been 'young and idealistic'. Then it all fell through, as there was a lot of in-fighting between the guerrilla leaders who in fact won the war and ended up being in charge in Addis Ababa and forming a government, but then fell out among themselves.

When you got back to the UK, I imagine it was a bit of a media circus?

It was terrible – we almost wished we were back in Addis Ababa. We don't like the bright lights, or consumerism – we just wanted to get back to Africa. We had a bit of time at home, and the good old British government took responsibility and gave us some compensation. Lindsay negotiated with the *Daily Mail* when he was in Khartoum for our story; we ended up with quite a nice little nest egg and we bought our cottage with it – so our cottage is really thanks to that episode.

Did you know that your parents knew you were safe?

From May to September, till John Swain came, we heard nothing from the outside world. They listened to the BBC World Service, which we caught snippets of, and that's when we heard about the $1 million. My dad had already died when we were in Ethiopia. We heard my sister, who's a doctor, on the radio offering to go out and provide medical services for the guerrillas in exchange for our release.

I read that you had to rescue your kids from a burning tent. Is that true?

I was in the Awash National Park, where I had gone with some friends as Lindsay was working. I stupidly put a candle in a plastic mug on a plastic case in the tent and we were just sitting round the fire enjoying a glass of wine when we suddenly realised that the whole of the tent was ablaze. It was frighteningly instantaneous, but we pulled the kids out and they were fine.

I read that you also had to bash some man over the head …

Yes, that was at Ambo Agricultural Station, an hour's drive from Addis Ababa. I was looking at some Mountain Wagtails by the river, wearing perhaps too short a dress, when some local man decided he was going to rape me, but we fought and I got away.

I think you then ended up with a broken leg?

I used to go riding with a friend. One day the horse bolted, I fell off and broke my ankle, and we had to go back on the one horse. My leg was set at the leprosy hospital by the Dutch doctors, giving me Dutch gin to keep me quiet while they yanked my bones back into position. You can go ringing with a broken ankle, I discovered!

Among the ornithologists in Britain at that time there were very few women, and I'd say you were probably one of the top one or two women?

When I went to Cambridge there was Daphne Watson, who was two years ahead of me. She was the only woman I knew at that time who was into birds.

She lives on the Isle of Wight and was very involved with the Merseyside Ringing Group. There were very few female ringers at that time. Now I'm delighted that it's changed hugely.

When you came back to the UK, did you go straight to Wales?

We rented a cottage outside Ross-on-Wye while we looked for somewhere to buy.

Is this where your interest in Dippers started?

Funnily enough, I got into Dippers when we were in Wiltshire and we lived near the Bybrook near Castle Combe. I was looking for Grey Wagtails but realised it was a lot easier to find Dipper nests, but then we went off to Africa and it wasn't until I came to south Wales in 1977 that I realised that the River Monnow, a tributary of the River Wye which rises in the Black Mountains, is just a wonderful river for Dippers, so ended up studying them and Grey Wagtails.

Did you have a job at that point?

We bought the cottage, then went to Tanzania for a year, and when we came back I got the Conservation and Development Officer job with the Gwent Wildlife Trust from 1979 to 1984. I'd done quite a lot of stuff with the Trust in a voluntary capacity, so when they advertised the position I went for it and got the job. I was quite happy doing the job, but it was exhausting as I didn't have any spare time – the job took up evenings and weekends and trying to juggle that with the kids was a bit of a nightmare. Lindsay was doing short-term contracts at that point, so he was away a lot. Then Roger Lovegrove came to see me and said the RSPB Conservation Officer post in Wales was coming up and why didn't I apply. That was a bolt out of the blue as I was very settled and the kids were at school in Monmouth, so it was out of my comfort zone applying for the RSPB job and moving to Newtown. A good move, though.

So you moved to Newtown?

The office was in Newtown, but I'd come back here at the weekends. The kids were in their teens so I was juggling car runs and keeping them on the straight and narrow as well.

You did much of the work on Dippers with Steve Ormerod. Is that because he was working on Dippers already?

No, Pete Ferns invited me to give a lecture in Cardiff. Steve was at that lecture – he was into invertebrates and we just got together and worked jointly for the next ten years. The projects extended way beyond the Black Mountains and we were covering huge areas of mid and west Wales. We got funding from the Central Electricity Generating Board to look at the impact of acidification on Dipper breeding performance.

Some of the published papers have you as lead author, sometimes Steve – how did you decide?

I think we just took it in turns. Steve is brilliant at statistical analysis, I like collecting the data, and we worked together very well.

You obviously decided to write the book (*The Dippers*), and you're the lead author?

I got a leave of absence to go and join Lindsay in Kenya, and every morning I worked away at it when we were in Nairobi.

So have you actually seen all the kinds of Dipper in the world?

Yes. I went with Steve to Nepal and we saw the Brown Dipper and various races of our Dipper, then went to Ecuador and saw the White-capped Dipper. Lindsay and I went for three consecutive years to Argentina and Bolivia to look at the Rufous-throated Dipper and we've also been to North America to look at the American Dipper – there are only five dippers, sadly.

Which is your favourite among those five?

Most definitely our own, but the South American ones are really smart and they're so different – they don't dive like ours do; they just spend their time on the rocks. As far as we know, they don't go underwater.

Are there any other differences that mark them from the others?

They flick their wings rather than bob, and they've got white patches on the wings which they display.

If someone gave you the chance to travel to Nepal, for example, to study the Brown Dipper, would you take it?

I think I've probably done all that now. The Brown Dipper is so like our Dipper. We watched it on some of the mountain rivers in Nepal; it's a larger and tremendously strong bird. I'm quite happy carrying on with my population study in the Black Mountains because that's been going for so long now and it's getting more and more interesting, in that the brood sizes are changing, and clutches are laid earlier. Every year is different. We had an appalling year in 2012 with the high water levels, and I think that caused a very low survival level of the young birds. Lots of nests were washed out anyway, so it's going to be very interesting.

How much of your time is spent on research and how much on enjoyable birding?

I do endless surveys for the BTO, Wetland Bird Survey, BBS and Waterways Breeding Bird Survey, so spring is quite busy doing surveys for others, but I suppose two or three days a week between the end of February and June is spent on Dippers and Wagtails. I'm quite a keen ringer and am training lots of new ringers, and do a lot of boring old garden ringing, which keeps me on my toes and keeps me up to date with ageing and sexing.

What about the state of the Welsh countryside these days?

Wales is appalling. Most of it's just a green desert, heavily agriculturally improved and overstocked with sheep. Most of the land is pastoral with very little ornithological interest left. Farmland birds have probably declined here as much as anywhere: Lapwings have gone, Curlew have virtually gone, Yellowhammers have all but gone. So many of the upland areas have also been improved and turned into grass moorland, and conifer forestation has also taken its toll, so it's pretty grim really. My big passion is recreating flower-rich meadows for bees, butterflies and so on – I'm Chairman of the Monmouthshire Meadows Group, which takes up a lot of time. We've got about 160 members, and we're advising them, helping to restore grassland, distributing seeds, working to try and get back some diversity, at least in Monmouthshire. Here we're not so badly off as some of the lowlands, in that we have got some very steep terrain and lots of little fields have escaped being agriculturally improved, so we've got pockets of land with species richness. I am also joint vice county recorder for the Botanical Society and again that keeps me busy.

We haven't touched on Botswana, with which you have a strong connection. What took you out there?

Lindsay and I drifted apart. He'd gone his own way doing shorter- and longer-term contracts and I'd been in Wales with the RSPB and my connections, working with Steve and so on, and we decided for the good of our marriage that something had to change, so we thought, how about going back to Africa? And Botswana came up. We went out there for three years initially and it was just brilliant. It's such a lovely country, very safe, and being semi-desert many of the birds are nomadic and things change from year to year. It doesn't have any endemics, though, but it's right next to South Africa, Namibia and Zimbabwe, so we got to know southern Africa as well. When the three years came to an end Lindsay was given the chance of extending for two years, which we did very willingly.

What period are you talking about here?

We went out at the beginning of 1996 and the contract lasted till the end of 2000. We stayed on for about four months and came home to the foot and mouth outbreak. We couldn't get out into the countryside here so we went back to Botswana – we had a vehicle out there with a roof tent that we lived in and I went back for six months in 2003. We were doing some work on Slaty Egrets. The Okavango Delta was then in the process of becoming a Ramsar site, identifying it to be of international importance for birds, and we had to help do a management plan. The Slaty Egret was one of the key species because Botswana has the bulk of the population. I went out for six months and worked on Slaty Egrets and I've kept that going, including convening a Slaty Egret workshop for the Africa–Eurasian Migratory Waterbird Agreement (AEWA) and writing the species action plan.

Who influenced you in later life?

In my twenties and thirties, John Ash and Michael Gallagher. Some of the people at the Wildlife Trusts, for example Ted Smith in Lincolnshire. Some of the amazing botanical people.

If you could meet someone from the ornithological world you've never met, who would it be?

I would love to have met people such as naturalist Joseph Banks, who accompanied Captain Cook on his expeditions and wrote and sketched all that he saw. It must have been so exciting at that time when so much was still to be discovered.

Had you not got into the line of work you went into, what would you have gone into instead?

There was no alternative. It had to be to do with birds, plants, conservation.

What's the best place in the world you've been to?

I loved northern Argentina and southern Bolivia.

If I could give you a ticket to somewhere in the world you haven't been to, where would it be?

Peru or the Amazon.

If you had to choose between the Dipper or Grey Wagtail to see for your last five minutes, which would it be?

Can't I see them both together on the same rock or under the same bridge?!

Is there a bird family in the world you like more than any other?

I love skimmers and cranes.

And what about the species you would most like to see?

That would have to be the Diademed Plover or perhaps the Ibisbill (which I have dipped out on twice).

What piece of music would you take to a desert island?

It would have to be a Mozart piano concerto or Jacqueline du Pré playing Elgar's cello concerto – wonderful, sad but uplifting.

Do you have a favourite film?

No, I hardly ever watch films, and can never remember what I have seen anyway.

TV programme?

Don't watch much TV either, really. I read and sit at my computer or I'm out in the garden or elsewhere, but I must admit I do like detective series – anything from Poirot to *New Tricks*.

Choose a bird book to take to a desert island ...

The complete set of *HBW*. Or, if not, one volume of *HBW* to enjoy the photos and at last have time to read the text. Failing that, the *Roberts Bird Guide*.

And a non-bird book?

How about a good atlas so that I can imagine being elsewhere? I love maps anyway, and can learn all the mountain ranges and rivers. I'll never have a sat nav in a car as I enjoy reading maps.

In the decades you've been working, how have you seen ornithology change?

Things have moved on hugely since the 1950s when all the fields were being ploughed up, and in the 1950s and early 1960s you couldn't imagine the groundswell of public support for conservation and birding that there is now. But at the same time we're still losing things, and although we've got that problem in this country and most of the European countries, in Africa and South America species are being lost as we speak. I just wish we could have a fraction of this country's manpower, intellect and knowledge diverted to Africa. You go to Botswana and there's a handful of people working in conservation and bird research, whereas here we've got thousands. I also worry about Asia and the decimation of wildlife there (and increasingly in Africa) because of the demand for snakes and other reptiles, birds and mammals, dead or alive, by the Chinese.

MARK AVERY

Dr Mark Avery is a blogger and author. He was born in the 1950s.

INTERVIEWED BY KEITH BETTON

At what age did you start being interested in birds?

It's quite difficult to remember – maybe about the age of seven.

Who fostered your interest in birds?

I used to go out for walks with Dad from an early age and he would point out easy birds like Kestrels, Green Woodpeckers and Magpies. I think those walks sparked an interest which burst into flame when I went to Bristol Grammar School (BGS) at the age of eleven. There was a very active Field Club, run by masters Derek Lucas and Tony Warren, and there were lots of keen birders amongst my fellow pupils.

Bristol was full of birders then as it is now.

First pair of binoculars?

I think they were a Christmas present when I was eleven – a pair of 8×30 East German things. Then a bit later I took over a pair of Swift 10×50 that my dad used now and again. And then my eighteenth-birthday present was a pair of Zeiss Dialyt 10×40B which I still have, thirty-eight years later.

I've seen every bird species I've ever seen through those although, for a few years, there was one species that I hadn't seen through them. When I was studying Bee-eaters in the Camargue I was sitting in the car, taking a blood sample from a Bee-eater's brachial vein, when a small flock of Rose-coloured Starlings flew past. I'd never seen them before but my hands were full of Bee-eater and a syringe so I couldn't grab the bins – it was quite a few years before my binoculars 'saw' that species.

What about a telescope?

I don't have one now because I can't be bothered with the hassle. That means that if I am out birding with you I might parasitise your telescope but I'm hardly ever bothered. If something is too far away to identify immediately, I just spend longer looking at it, try to get a bit closer, or shrug my shoulders and go and look at something else. If I did more sea-watching then it might be different – but we don't have any coast in Northamptonshire. Come to that, we don't have that many birds quite a lot of the time!

I did have one of those old brass Broadhurst–Clarkson scopes when I was at school – it fell apart eventually!

Were you in the YOC?

Yes, even before I went to BGS, so probably from the age of eight or so. I remember sending in a sighting of a Hoopoe that I had seen with my dad near Slapton Ley in Devon on an Easter holiday – but I must have seen that without binoculars and without having a field guide either. Luckily, they are unmistakable!

The YOC wasn't that important to me: there wasn't a local group that I attended, but I think it helped to inform me about birds and it certainly showed me that I wasn't alone in being interested in birds.

Books?

I remember taking *The Observer's Book of Birds* out on car trips, and loving thumbing through the *AA Reader's Digest Book of British Birds* in the evenings. I think Malcolm Ogilvie's Poyser, *Ducks of Britain and Europe* was the first one of those that I bought, but the Peterson, Mountfort and Hollom field guide was the most important book because it opened up the world of birds in Europe and was such a good identification guide. I've had three of those – I wore out two – and my current copy feels like an old friend, even if I don't look at it much these days.

John Gooders's *Where to Watch Birds* was very important too – in those days, before the first *Atlas* (by Tim Sharrock) was published in 1976, it wasn't easy to know where to go birding if you were in a distant part of the country. I knew my local birding haunts but if I had been put down in Scotland I would have been lost without John Gooders.

It sounds as though from an early age you went birding at least once a week?

Yes, I think so. In the holidays I was always going out for walks in the north Somerset countryside, south of Bristol, or cycling to Chew Valley Lake, about six miles away, for the day, and visiting all the hides.

When I went to university the amount of birding I did definitely dipped for many years and then it came back in my thirties. Nowadays I visit my local patch of Stanwick Lakes at least once a week and I do BBS surveys, and try to fit in a bit of birding to any work that I am doing away from home. I'm still a birder but I'm very happy for most of my birding to be local to home. After all, there are birds everywhere.

Did people think you were a bit cissy, as a child, being interested in birds?

I think I had more of a problem being the only boy from my village who went into Bristol to the posh grammar school, in terms of name-calling and peer pressure.

I'm an only child and that might mean that I am self-contained (it might mean other things too), but I had friends locally that I played football and cricket with, but I could always go for a walk and look at nature if I was at a loose end.

Did you ever think you would end up working on birds?

No! Not until I did end up doing that. I should have taken more notice of my biology teacher in the sixth form, Ron Cockitt, who said to me that I'd probably end up working for the RSPB. I really didn't imagine that might happen, and it certainly wasn't a career goal – but then I've never really had one of those at any stage; I've just done the next thing that looked most fun in life.

Were family holidays important?

Yes, very. I got to see new birds! And that's where John Gooders's book helped a lot. I suspect – in fact, I know – that I probably nagged my parents about where we might go on holiday, but they were both interested in the countryside and walking and scenery so I didn't have too hard a job influencing our holiday destinations now and again.

When we first went to East Anglia, I think I was only twelve or thirteen. We went to the Ouse Washes, Cley and Minsmere, and it was simply fantastic. Those were some of the first nature reserves I had been to which were large and had a list of birds seen that day that one could examine before rushing out to try to see those species. That holiday produced my first Avocet, first Bittern, first Marsh Harrier, first Bearded Tit – 'firsts' of lots of things.

My first Turtle Dove was near Wicken Fen, and I remember thinking how beautiful a bird it was. After that first sighting we saw them all the time, every day. It's not like that now. I've been back to that very site and not seen a single Turtle Dove.

How about foreign holidays?

I went on a French exchange to Brittany, when I saw my first Serin, but we didn't travel abroad as a family. As an undergraduate I went on an expedition to Norway ringing waders – Purple Sandpipers, Dotterel and anything else we could get our hands on. I saw my first Long-tailed Skua, Common Crane, Bluethroat and Lapland Bunting on that trip, but not Snowy Owl – that's a bird I still haven't seen.

Did you keep a notebook?

As a kid, yes, but it didn't have anything very interesting in it. I haven't got any of those old notebooks, which I slightly regret, but would I do anything with them? Actually, what I would do is put the bird lists into Birdtrack now.

Birdtrack is my notebook these days – at least, for recording species lists. I like the fact that I can look up the last time I saw a Yellow-legged Gull, a Spotted Flycatcher or a Kingfisher at Stanwick Lakes, and I also like the fact that my casual birding observations are there to form part of a national picture of what is happening to birds: numbers, ranges, arrival and departure dates, etc. I like the fact that Birdtrack is useful to me as a birder, but also that my data may be useful to others in future. That's much better than a pile of notebooks that would be thrown away after I die.

Might any of your early birds be changed now because they weren't what you thought they were at the time?

I think we could dismiss the Red-backed Shrikes I saw at Lulsgate Airport – they were Linnets, of all things (but I think I was eleven then, and probably was working without binoculars and without a proper field guide). And I thought I had seen a Jack Snipe for ages before I actually did. But from the age of thirteen or fourteen, when I got through the 'immensely keen and wildly optimistic' stage, I was pretty reliable, I would say.

Have you found any rare birds?

Not many – I wonder what I've missed! While holidaying in north Norfolk, in 1974, two friends and I saw a Black-winged Pratincole on Salthouse Heath in late August. We had spent all day birding on the coast and there wasn't much about so we went onto the Heath just for something to do. This bird flew over and I remember thinking, 'Is that a wader? No. Is it a tern? No. I've never seen that before. Oh, I know! I've seen it in Peterson, Mountfort and Hollom, it's a pratincole of some sort,' and then my friend, Peter Dolton, said, 'Pratincole!'

We saw it again, a day or two later, in the evening, at Cley and then there were a few other people too. I ran from the hide to get the warden, Billy Bishop, who came and checked our sighting. He gave us a cup of tea and showed us a drawing that Richard Richardson had done of a previous Black-winged Prat at Cley many years before. I knew, even then, that Richard Richardson and Billy Bishop didn't have an easy relationship, so I was interested to see the work of one in the home of the other.

But, no, I haven't seen many rare birds.

Were you ever a twitcher?

Not really. I probably had the mentality but not the means when I was in my late teens, but I didn't drive, didn't have any older friends who did drive, and then I went to university for three years, then spent a year working for John Krebs at Oxford, then three years doing a PhD and then a couple of years doing a post-doctoral research fellowship back at Oxford, so I was busy studying birds by then. And I was much more interested in how birds had evolved than what misplaced bird was on the Norfolk coast.

I'll tell you one story, though. In spring 1975 a junco, what we then called a Slate-coloured Junco, turned up in the Cotswolds, in Gloucestershire. It wasn't a million miles from home but I was dependent on my kind parents to take me, and a couple of friends, to have a look for it. It was at a place called Haresfield. We got to this village and there wasn't another birder in sight – everyone had been and gone over the ten days the bird had been present.

The only person we could see was a guy cutting his lawn so my mum said, 'Go and ask him,' but I was too shy to ask. Mum said something along the lines of, 'Well, I'm not coming all this way just to turn round and go home again,' so she asked the guy if he knew anything of a rare bird, and he said, 'Yes, it's in my back garden. Do you want to see it?', to which Mum replied that she wasn't the least bit interested but there were three boys in the back of the car who were dead keen.

So we saw it. I'm less shy now.

Although you didn't think of the RSPB as a job, did you think of a job with birds?

I kind of fell into it, really. As an undergraduate, I met a friend, from BGS, by accident, in the Eagle in Ben'et Street. He asked what I was doing in the summer vacation, I said I didn't know, so he suggested working on the Red Deer Project on Rhum in the Hebrides. I was lucky enough to spend the summer with Tim Clutton-Brock (now Professor Clutton-Brock, FRS) and his team studying Red Deer. That changed my life, because the talk around the dinner table, on the northernmost point of Rhum, was of selfish genes and kin selection and how evolution shaped the behaviour that we were seeing every day. When I went back to Cambridge for my second year, I switched all my options from biochemistry, which I had thought I would do, to animal behaviour and evolution. That changed my life.

Did you think you were happy before you made that change?

Yes, completely happy! How could I not be happy? I was away from home for the first time (not that there was anything wrong with home); I had a grant (which didn't turn into a debt like these days); I was in Cambridge, one of the best and swankiest universities in the world; beer was about 20p a pint in the college bar; I went to an Indian restaurant for the first time in my life; I loved learning about biology; and I had been to an all-boys school so there were many new things to explore. Who wouldn't be happy?

My children have both been through university now (my son is doing a PhD) and my advice to them was to enjoy the three years – but don't waste them. I guess my motto would be something like 'work hard and play hard', and that's what I did at university, I did work hard academically but I had a lot of fun too: going on marches against apartheid, going to punk concerts, playing squash instead of going to geology practicals, attending Cambridge Bird Club meetings, making friends, drinking too much sometimes, going on an expedition, going to Rhum. Just fantastic!

But did you know what you wanted to do in life?

After Rhum, and even more so after the expedition to Norway, I knew I wanted to study animal behaviour, preferably birds.

When I finished at Cambridge I applied for some PhD positions and was offered one that I turned down, and failed to be offered several that I quite fancied. But I was offered a job by John Krebs (now Professor Lord Krebs of Wytham Woods, FRS) as his research assistant. That was lucky too.

John had heard me give a talk at the 1979 EGI student conference at Oxford about the work I had done (with Geoff Sherwood) on Great Snipe lekking whilst in Norway. I don't think it was a very good talk, but it was fairly unusual for an undergraduate to give a talk and there must have been something about me that caught his eye. John came up and talked to me at the coffee break and was very nice, but I didn't really know who he was (I'd heard of his father, who had won the Nobel Prize for biochemistry).

Anyway, John offered me a job and I remember phoning him from Liverpool – I could show you the spot at Lime Street Station – after I had turned down a PhD there.

I spent a year working for John in 1979/80 and then went back to Oxford as a Research Fellow in 1984 and 1985. It was a great time to be in Oxford if you were interested in birds and evolutionary biology. Richard Dawkins, John Krebs, Chris Perrins, Bill Hamilton and others were there, and a bunch of research students and post-docs like myself. I learned an awful lot, and had an awful lot of fun.

John was a great guy to work for. He's very clever and I learned a lot from him: a lot about how to think as a scientist and how to write scientific papers, and also a lot about management and leadership. He was a great boss. I've been lucky with my bosses, and I owe him a lot. We were studying Great Tits and Marsh Tits in and around Oxford, and Bee-eaters in the Camargue, so that was all good fun too.

So I did end up working on birds, as a scientist. But in the period 1981–3 I did my PhD, based at Aberdeen University, studying Pipistrelle Bats. More fun!

How did you get from Oxford to the RSPB?

Another chance event! I got a short contract with the Forestry Commission (FC) at the end of my time at Oxford, after my fellowship money had come to an end, and I was looking for jobs, and the job was to help Steve Petty of the FC write a review about woodland birds.

In doing that I came across a draft paper by Colin Bibby, the Head of Research at the RSPB, which I read, and I thought I found a mistake in it so I phoned Colin and he said I ought to come over for a chat. A few days later, I was sitting in Colin's office at The Lodge (not realising that one day it would be my office and one day, his job would be my job), and we talked about his paper. He had made a mistake.

What was it?

I can't exactly remember. It was a logical error in the analysis, quite important to the results, but a bit abstruse. Anyway, Colin was pleased to have this pointed out to him because it meant he could re-do the analysis and change the paper before it was published. And then he asked me whether I'd like to work for the RSPB in the Flow Country leading a team studying the impact of afforestation on moorland birds. That wasn't exactly what I had expected.

I phoned him the next day and said, 'Yes, please.'

Quite often I am asked to give advice to young people wanting to get into nature conservation, and I'm really not much help when I say go to the pub (and get offered a summer job) and find mistakes in people's work (and hope they offer you a job).

About five years later when Colin left to work for BirdLife International, I got his job as Head of Research and then seven years later, when Barbara Young left the RSPB for English Nature, and Graham Wynne became Chief Executive (having been Conservation Director), Graham appointed me as Conservation Director, in which role I worked for Graham for another twelve years and then for Mike Clarke for a year until I left in spring 2011.

To what extent did you influence Graham Wynne, and to what extent did he influence you?

Well, quite a lot in both directions, I think, since we worked together for about nineteen years. But he was definitely the boss!

I learned a huge amount from Graham – not much about birds, but lots about management, and leadership, and political advocacy and just 'stuff', because he is a bright guy with lots of experience and a quick brain. I said I'd been lucky with my bosses – by the way, Colin Bibby was a great boss too.

One of the great things about working for Graham was that he was very clear about what we were trying to achieve. He was clear about the outcomes he wanted and he was relaxed, but also had a view on how to get those outcomes. And when some of what you are trying to achieve is through political advocacy, sometimes sitting in private with a government minister, sometimes talking on the *Today* programme with a few million people listening, sometimes in some heated discussion with the shooting community or the National Farmers' Union (NFU), it really did help if you knew what your boss would support and what he wouldn't. You are speaking for the organisation and you have to make decisions on the spur of the moment, so that certainty helps an awful lot. So, I hope, I was reasonably good at giving that clarity to my staff too, because they were often in a similar position.

Were there any occasions when you got it wrong and Graham took you to task?

There were plenty of times I got things wrong! I'd often know when I got things wrong and go and admit it, and Graham and I used, quite often, to have chats in the evening if we were both in The Lodge – we were often the last people there – about what was going well and what was going badly. But Graham was clever enough more often to say something like, 'I'm not sure that was absolutely the best thing to do, Mark, but we'll see,' rather than take me to task – he knew that would have much more of an impact on me than telling me I'd got it wrong.

I think I remember you saying to me once that you wrote quite a lot of things that went out under Graham's name and so you knew how his mind worked really well, and it was something like a marriage?

I'm not sure it was that much like a marriage! But we worked closely together over a long period of time and we did know each other's thoughts and opinions very well. I did draft the Chief Executive column in *Birds* (now *Nature's Home*) magazine for over a decade, but Graham and I would go through them together, often late into the evening on the last possible day before the *Birds* magazine team needed the finished copy, so it was definitely Graham's thoughts that went out. But it was my job to do most of the work to get to that position.

There's probably one area where I think you weren't on the same wavelength: the persecution of birds of prey. My perception is that you felt more strongly than Graham that this was something that the RSPB should take a stand on.

I did feel that a bird conservation organisation should take a strong stance on people killing birds of prey illegally, yes. But I don't think Graham thought differently. And the RSPB policy, set by the RSPB Council, certainly didn't leave us in any doubt about where the RSPB should stand.

But I think this was an area towards which, as a keen birder, I naturally gravitated. Graham was a town planner by training and I found the planning system difficult to understand and tended to leave that area to him as his strength. I think Graham was happy to let me take more of a lead on raptor issues because they suited me. Just as he would know if someone were talking rubbish on planning, and I would struggle, he knew I would be a better advocate on predator–prey relationships and things like that. Putting Graham in a room full of gamekeepers and me in a room full of planners would be crazy if the other, or somebody else better, could do it!

You spent a long time at the RSPB, much of the time as Conservation Director, but you didn't become Chief Executive. Do you think, looking back, that you should have left earlier and perhaps become Chief Executive somewhere else?

I don't have any regrets. I had a great time at the RSPB and worked with a great team of people. I could have left the RSPB at various stages as jobs were dangled under my nose, but I wanted to stay at the RSPB because I thought then that it was the best nature conservation organisation in the UK. And I still do think that. Working for the RSPB was like playing for Man. United – of course, Man U. have slipped down the league a bit recently, but the RSPB is still at the top, I would say.

What do you think your external reputation was when you were at the RSPB? It could be said that there were some groups that you didn't get on with as well as you might have done. How do you think you were viewed?

It's slightly difficult to tell because it's very difficult to tease apart whether a group's opinion is driven by the organisation for which you work or by what they think of you, personally. In advocacy terms you have to decide whether the people who are stopping you getting your way (the way that would be best for wildlife, that is) can be persuaded by reason or charm – or not. If you can persuade people to farm better for wildlife or stop killing Hen Harriers by smiling at them and telling them how great they are then that's what you should do. If you think that that friendly approach is unlikely to work then you have to treat them as a problem and find a way to enforce the law or adjust the economics of the situation so that they change their behaviour.

With the leadership of the National Farmers' Union and with grouse moor managers, I decided that the latter course ought to be the one we adopted. I still think that was right. Not because we had huge successes, although we had some successes in agriculture (though not enough of them, and largely *despite* the National Farmers' Union rather than because of it), but because we would certainly fail if we pretended that there wasn't much of a problem and that everyone was doing their utmost.

So, if I had believed that a different approach would have yielded more birds on the ground, I would have taken that approach. However, there is, I think, a good reason for speaking out if in doubt, and that is that as an NGO you depend on public support. You have to stand for something. And one of the things that the RSPB had to stand for was that birds should not be killed illegally, while another was that the decline in farmland birds is unacceptable.

There is quite a lot of pressure not to stir things up. The *Shooting Times* seemed to have something every week slagging off the RSPB, and me in particular, but you grow a thick skin. Talking to the National Farmers' Union and moorland managers, and expecting a sympathetic hearing, is the UK equivalent of thinking that you can have a reasonable chat with a Maltese hunter about not slaughtering birds. The issues are different, but the views are similarly entrenched. We aren't all on the same side and it would be madness to think that we are.

On a personal basis, I actually got on really quite well with many grouse moor owners and many, many famers. There are lots of people with whom I have argued in the media and in front of a government minister but with whom I would be happy to share a beer or a bottle of wine. Indeed, I have shared more than a bottle of wine with some of them since leaving the RSPB, and we have joyously fallen back into the same debates and arguments. I find some of 'the enemy' much more engaging personally than I find some of 'our own side'.

Some of those people loathed me, but I didn't mind. Some thought I was unfair to them, and some thought that I was quite a good adversary and they couldn't help but like me personally. And that's how I felt, and feel, about them too.

How difficult was it to handle the situation in autumn 2007 in which the RSPB's patron's grandson was in the press because he had been nearby when two Hen Harriers had been reported as being killed on the Sandringham Estate?

Quite tricky, but not in the way I think you mean. Not tricky because Prince Harry was involved and the Queen is the RSPB's patron – that didn't make it tricky. But because it was a media frenzy for a few days we worked very hard to put our view across. And our view was: this was a serious incident that should be investigated. If the evidence was sufficient, then a prosecution should be brought. We would help the police in any way we could; there is too much illegal killing of Hen Harriers going on, and we were against that.

The funny thing is, we were accused of both stirring it up and remaining completely silent on the matter, two views which are difficult to reconcile. I was more upset by those who said the RSPB had kept a craven silence about the matter, because they tended to be our friends, and they were wrong. I knew they were wrong because I did little else for days than do TV and radio interviews and talk to the press.

On a lighter note, what is your favourite film?

Once Upon a Time in the West, or maybe *Four Weddings and a Funeral*.

Your favourite TV programme?

Antiques Roadshow, not because I have any interest in antiques but because I like to hear experts and enthusiasts talking with enthusiasm.

And your favourite music?

Either Maria Callas singing *Tosca* or Bruce Springsteen's 'Born to Run'.

What about your favourite book?

Hemingway's *For Whom the Bell Tolls*.

If I were to give you the power to go to the government and change three things in the world of conservation, what would they be?

The first would be that an absolute cornerstone of nature conservation is having a network of protected areas – and we still don't have that for the marine environment. We need the best areas for wintering seaducks, the best feeding sites for seabirds, the best sites for cetaceans, and a whole load of fish spawning sites to be properly protected.

The second is that we need to make the £3 billion per annum that you and I (as taxpayers) give to farmers deliver a better environment. The RSPB's Hope Farm has shown that you can produce lots of food and lots more birds than the average farm very easily. I want that to happen as part of my dividend in return for investing so much of my taxes in farming. That would mean we could have more Turtle Doves back in the Fens and more Skylarks and some Grey Partridges on my BBS square in east Northamptonshire.

The third would be to ban driven grouse shooting. I don't like banning things – but I think that banning grouse shooting outright might be the only way to get the Hen Harrier back to its former status in the UK. And grouse moor management is environmentally harmful in lots of other ways too. Don't get me started!

A fourth thing that I would like to see, which affects the government but isn't an 'ask' of the government, is for birders to become more active as campaigners themselves. Our wildlife conservation organisations are pretty good – though, as I've said, I think they are being rather too quiet about the state of affairs at the moment. However, there is an awful lot that we can do as individuals that doesn't rely on organisations to represent our views. My column in *Birdwatch*, 'The Political Birder', sets out things we can all do to influence the world.

I'd like to see more birders on Twitter and Facebook campaigning for a better environment. I'd encourage all birders to contact their political representatives regularly about environmental issues too. Visit internet sites like Avaaz (www.avaaz.org/en/) and 38 Degrees (www.38degrees.org.uk/) to see what environmental campaigns they are running, and get involved. Read blogs like that at Raptor Persecution Scotland at http://raptorpersecutionscotland.wordpress.com/ and my daily blog 'Standing up for Nature' at www.markavery.info, and make your views known. If you do it alone it won't make much difference, but if we all do it together then it most certainly will. Birders of the UK unite! You are losing your Hen Harriers, Corn Buntings and Spotted Flycatchers. Raise your voices to get them back! They can't speak out for themselves; they are relying on you.

STEPHEN MOSS

Stephen Moss has been responsible for many of the best natural history TV programmes of our time. He is also a prolific author. He was born in the 1960s.

INTERVIEWED BY KEITH BETTON

What's your first memory of birding?

My first memory is before I can remember, if that makes sense. My late mother used to tell me the story – which has become part of family mythology – that she took me down to the River Thames at Laleham when I was two or three years old, and said, 'We're going to feed the ducks' – something every child in Britain has done at some point. And I said, 'What are those funny black ducks?' and she said, 'I don't know, dear, but we've got a little book at home that one of your aunts has given you [of course it was *The Observer's Book of Birds*]'. I was a precocious little thing then and I went back, looked it up and realised they were Coots. Then I basically read the book from cover to cover and memorised it – I can still remember it starts with Magpie and ends with Capercaillie. That really got me going, and I can't remember really not being interested in birds.

So did you carry on birding from then?

At school I did projects on birds when I was six or seven; I was very lucky to be of that generation who were 'allowed out' to play. I had a very worried grandmother – who brought me up with my mother – who didn't want to let me out, but from the age of five or six we would bunk over the back fence to look at birds. Then from the age of about eight I went to the local gravel pits at Shepperton, and I suppose the bird that really sums it up for me is the Great Crested Grebe because in the third year of primary school, aged about nine, we went on a nature walk and I remember wanting to see Great Crested Grebe and being aware of this bird in my peripheral vision. I looked to my left and there was a Great Crested Grebe sailing along. I thought it was an extraordinarily beautiful bird, and it's still one of my favourites.

We were a single parent family and my mother didn't take any interest in birds but she would take me on holidays to the New Forest, Milford-on-Sea, Keyhaven Marshes – I remember vividly seeing my first Blackcap and Stonechat at Keyhaven when I was ten. Later we went to Minsmere, where I met the great Bert Axel at the age of thirteen. Mum even took me out of school during my O-Levels year to go to Scilly in autumn. So that's what really got me going, and I owe a huge amount to my mother.

What were your first binoculars?

I was nine years old and my mother's boss lent me a very expensive pair of Zeiss binoculars to go to Staines Reservoirs for the first time. That year, my mother bought me a pair of Wallace and Heaton binoculars for Christmas – they were £14.9s.6d. When I was thirteen, on the way to Minsmere, I got a pair of Carl Zeiss Dekarem, those huge binoculars, which were wonderful, and I used them to see my first Avocet.

What about your first telescope – how old were you and what was it?

I got a little Greenkat spotting scope when I was about fifteen and went to Stodmarsh with my friend Daniel. I left it somewhere, realised, ran back after about five minutes and someone had pinched it. I didn't get another one till I was in my early twenties, when I got a Bushnell Spacemaster, and that did change my life.

What came after the Bushnell?

I think I went straight to Kowa in the mid-1980s, and I've stayed with them ever since.

What about binoculars?

After the Zeiss I got a pair of Zeiss 10×40, then 7×42 Dialyts in the 1980s, and now I have a pair of Leica, a pair of Swarowski, and a couple of Opticrons.

Did you join the YOC?

Yes, my first trip was to Staines Reservoirs on 17 November 1969 when I thought I was going to die of exposure. I was wearing a pair of thin trousers, a shirt, a woolly jumper and a not very good coat, and it was absolutely freezing. I remember seeing a Cormorant flying over but missing a lot of good things because I wasn't very aware. I enjoyed *Bird Life* magazine but I wasn't a great 'joiner' and didn't go on YOC holidays, which I really regret now as I've met people who did and who really enjoyed them.

Did you hand in any 'special notes' to *Bird Life* magazine?

There were some wonderful notes. There was a chap in Cornwall who shall remain nameless who kept sending in notes like 'Stumbled across Little Bittern round the back of my house'. I remember even then at the age of ten thinking 'Stringer', as he clearly was. I did send a note into *British Birds* when I was fifteen: 'Starling imitating Cetti's Warbler', and there was a letter in *British Birds* afterwards suggesting what we'd heard was in fact a Cetti's Warbler. I have a horrible feeling that the writer might have been correct, but I'm not going to write in and correct it at this late stage!

Talking of stringing, are there any things you know you've strung?

Endless things. I had Peterson's guide from the age of seven or eight because it came out in 1966. I had a Red-throated Pipit (which was a Song Thrush) around our house in Shepperton. I was absolutely convinced for years that I'd seen an Alpine Accentor outside my mother's front window – which I have to reluctantly admit was a Dunnock. My first Nutcracker was probably a Starling. Perhaps the best one was when I went on holiday to Milford in 1969 when I was nine and there was a flock of birds in the garden of our boarding house. I didn't know what they were, looked up the description in Peterson and decided they were Bar-tailed Desert Larks, the first record for Britain. By the following year I'd realised they were Linnets – but, to be fair, if you read the Bar-tailed Desert Lark description (as there was no picture), it's pretty close to a Linnet. But I didn't submit any of them, and I suppose it was the classic problem of having a book which covered all of Europe's birds.

Was Peterson your first bird book?

The Observer's Book of Birds, then Peterson, then of course the book I'm sure every family in Britain had, the AA Reader's Digest guide with the owl on the cover and Ray Harris-Ching's illustrations, which are extraordinary even now. I met him many years later and it was one of the great moments of my life. When he did the book he'd only just come over from New Zealand, and had never seen any of these birds, so he went to [the Natural History Museum at] Tring and got all the skins out – which is why they're all in very bizarre poses – but they're still incredible. That book inspired me because I wanted to see the birds I hadn't seen, such as Lesser Spotted Woodpecker, and I'd make lists of how many I'd seen in each section. A little bit later, in my teens, it was the Heinzel, Fitter and Parslow guide. About ten years ago I was walking round the Birdfair and someone said to me, 'Oh, Stephen, have you met Hermann Heinzel?' and I was like, 'You're kidding me?', but I met this extraordinary man who was in a Nazi concentration camp as a Jewish child and had got through the war, and produced this incredible guide. I assumed – as you often do with people from that generation – that he must be dead, but there he was, large as life. That was a seminal book for me.

I write books now about my love of birds and why birds are special, and I think during the period when I was growing up that Eric Hosking's book was the only one that did that. Guy Mountfort's books were earlier than that, of course, but I think Eric's books were the only ones that conveyed what it was like to go and look at birds, and weren't done in a field guide or scientific way; his enthusiasm shone through. People like myself and Mark Cocker owe a huge amount to people like him because there were very few books in those days that made you want to go to places and look for birds.

In those days, who were your main birding companions?

I heard about you in my first week at Hampton School, when my friend Andrew Chiverton said, 'You need to meet my friend Keith Betton, he's into birds.' I did meet you four or five years later on the causeway at Staines and we talked about waders. In my first week at Hampton Grammar School I sat next to a boy called Daniel Osorio and we got talking and realised we were both interested in birds. I think if I hadn't met Daniel I would have stopped and left it all behind and gone in for stamp collecting, because in those days you didn't meet other birders – you met all these rather arrogant and clever blokes on the causeway at Staines, but they were all in their thirties and forties and a bit terrifying.

Daniel came from this very bohemian Jewish family where he was the eldest of four brothers and sisters. Within a year or so of us knowing each other, and going birding in Bushy Park and Staines, his mother said, 'We're going on holiday to Norfolk for the October half-term – do you want to come? We've got a camper van, and we're staying with some friends, so frankly another child is neither here nor there.'

So I went, and for the next three or four years during my absolutely key years from thirteen to seventeen, Daniel's family effectively adopted me, knocked the edges off me as a spoiled only child, and took me to amazing places, such as the Scottish Borders, Dorset, and Norfolk several times, and we saw some fantastic birds. They let us hitch-hike, they let us cycle off, they believed in 'free-range children'. I dedicated *The Bumper Book of Nature* to them. When I told Daniel's mother this, she said, 'I thought we were just like other parents,' but my mother was terrified. They let Daniel and me go off cycling to the New Forest camping for a week when we were fourteen, and we kept forgetting to ring them (but we did see Ross's Gull!). So I owe them a huge amount. I still see Daniel; he's now a Professor of Biology at Sussex University.

At that time, apart from your family, who was influential for you in birding?

I wrote to Ian Wallace when I was thirteen because I'd been to Minsmere and seen what they thought was a Semipalmated Sandpiper, then was thought to be a Red-necked Stint and is now thought to be a Little Stint – and he wrote a very nice letter back. I knew about Jeffrey Boswall (who I later met when I worked at the BBC), Bruce Campbell, and obviously Peterson, Mountfort and Hollom. I suppose I never thought of them as real people I would ever get to meet, but I did. I used to see Eric Hosking at various *British Birds* presentations – I never plucked up the courage to go up and say hello, and I seriously regret that.

Anyone else you regret not having met?

Peter Scott, and I'd love to have met Roger Tory Peterson, which some of my friends did. I tried to interview Guy Mountfort for *A Bird in the Bush* but sadly he was too old by then. I did get to meet James Nicholson, and Richard Fitter and I became friends with Ian Wallace and James Ferguson-Lees, whom I consider to be the godfathers of post-war British birding.

Did you take photographs?

Not really. When I graduated from university my mother bought me a Vivitar zoom lens and I went to Shetland and took quite a few photos.

Sound recordings?

No.

Did you know Dominic Couzens in those days?

I only met Dominic fairly recently, about fifteen years ago, when a dear friend's parents, who are sadly both now dead, used to go with Dominic on his walks. One day Wendy said to me, 'You ought to meet my friend Dominic Couzens, you'd have a lot in common,' so she arranged a sort of 'blind date' and I went to Dominic's for dinner. We've been close friends ever since.

Notebooks?

I did keep notebooks but I've got to be honest: I was pretty hopeless. I've got lots of tick notebooks with some fairly dodgy sightings in, just things I thought

I'd seen but probably hadn't. I kept a few back in the 1970s and 1980s, and I've kept detailed computer notes since 1999. While I was filming with Bill Oddie I saw some good birds but I didn't take notes because I was working, so sometimes I'm not absolutely sure what I've seen in Britain as I'm not that interested in listing.

Did you ever think about working in ornithology?

I wanted to do biology or zoology at university – I wasn't very good at any of the other sciences but I was good at English. I remember getting a lift to school from a neighbour who chatted to me about my future and said, 'You're obviously very into birding, but studying biology is a very different thing, so why don't you keep birding as a hobby?' It was the best piece of advice I was ever given. I did an English degree at Cambridge, loved it, did some student journalism and got a job at the BBC. And for many years, although I worked briefly at the Natural History Unit when I was in my early twenties, I wanted to keep my job and hobby separate so I came back to London, had a family and got into writing about birds and then got the first *Birding with Bill Oddie* series off the ground. I then got the famous 'Do you want to play for Manchester United?' question and I was able to go and work for the Natural History Unit. I've never regretted it; it's been wonderful, but I'm rather glad I didn't spend my whole career in the birding area and had a career with the BBC for fifteen years making non-wildlife documentaries. And I'm really glad I didn't go to university and do zoology: I'd have been hopeless at it, and I wouldn't have enjoyed a career as a biologist.

Of all the birds you've seen, is there one that stands out above the others?

I have an interesting take on birds, and it's rather like Nick Hornby when he wrote *Fever Pitch* about supporting Arsenal: when he sees a player score a penalty in a certain way it reminds him in that instant of a player perhaps from the early 1970s. I feel that way about birds, so the Jackdaw is a big bird for me because there is a photo of me aged about eighteen months old (even before I was feeding the ducks and wondering what the 'funny black ducks' were) holding up my hand to a tame Jackdaw. Since then, I've seen Jackdaws in big flocks in Israel, in a country village on New Year's Day, and over my house at night when I lived in Hampton, so for me the continuity of a bird is important.

The older I get, the more I find that when I see a bird, it encapsulates all the other sightings I've had of that bird. For instance, I went looking for Cranes in Somerset the other day – we didn't see them, but I remembered the 20,000 cranes I filmed in the Hula Valley back in the late 1990s. And this is true of any bird. It could be special or very common, and as I get older it's more about what wildlife and nature mean to me, in that it reminds me of incidents from my childhood. The Great Crested Grebe would be my 'desert island bird'. So maybe the Great Crested Grebe, rather than the Jackdaw, is my memory of a really special bird. Another one is the Little Egret, which I first saw on Brownsea Island when I was ten and now I see them over my home, which reminds me of all the other times I have seen them.

I interviewed Mark Golley for *A Bird in the Bush* and I asked him at one point, completely unthinkingly, 'What would have happened if you'd never had that first encounter with birds and your life had gone in a different direction?' We both looked at each other, a chill ran down our spines, and I almost had tears in my eyes as I thought I couldn't imagine what life would have been like, which sounds rather dramatic. A friend of mine saw me chatting at work to a colleague who'd seen a Bullfinch in her garden and said, 'You're rather like one of those born-again evangelical Christians asking people if they've heard the "good news" about birds!' I like the fact that I can convert people, and if I can convert them when they're young – as my children are now – or even if they're much older, if they get even a tenth of the joy I've had out of birds, then that will be special to me.

Looking ahead, what are you going to be doing in five years' time?

I've had a lovely time making television programmes, but I may not do it again, as I'll be in my late 50s by then. I will always write about birds, and when I do it's a way of recording my experiences, of making sense of them and – I hope – inspiring other people. The thing I'm most proud of is *Birding with Bill Oddie* and *Springwatch* – I'm proud collectively that my colleagues and I did that, not just me. When I write a book and someone comes up to me and says, 'I really loved that passage', that's really important to me.

If something happened in ten years' time which meant you couldn't watch birds, what would you do instead?

I've thought very hard about, for instance, what if I went blind? Obviously, I could still listen to birds and I think it would open up a whole new way of life. I can't imagine ever not being able to watch birds in some form, though, and I think the interesting thing about birds is that no matter how old you are or what state of health you're in, you can still enjoy birds. I've just written a book about the birds and wildlife of my village. Most of my wildlife watching is done in my garden: I've seen twenty species of butterfly in my garden, a third of Britain's butterflies, and if I could stay in my village I'd enjoy it and make something of it.

Have you ever been into anything else, such as stamp collecting?

I love butterflies and dragonflies and am getting into moths, which I can still twitch! I've given up most of the things I did as a child, such as stamp collecting. My children are a very important focus for me, particularly the younger three, and getting them into birds is important to me. This morning I had the most wonderful thing with George running in off the hotel balcony and saying, 'There's a Bullfinch' – which of course we doubted, but there it was!

Let's move forward twenty years. What's birding going to look like?

I'm very worried about birds and birding. I wrote about weather and climate change in 1995. I'm very concerned that we'll have lost birds that we took for granted like Corn Buntings and House Sparrows, House Martins and Swifts – if Swifts didn't come back, that would be the end for me. We won't get so many young birders, as they won't see the point. I'm optimistic that this movement of (a) appreciating British birds and wildlife; and (b) appreciating it from an emotional human response, is going to continue to grow. We lost it in the 1960s and 1970s when everything had to be scientific and about identification and behaviour – nothing was about the human response – but it's coming back now. So I'm both pessimistic and optimistic.

What was your best day's birding?

In 1998 I was in Tobago at Buccoo Marsh with Suzanne, my wife to be. We had just got together and it was an amazing day with beautiful birds. I was going through a difficult time: my mother had died, my previous marriage had broken up, but this was a very special day. But I have a feeling that Buccoo Marsh has been drained and ploughed up – if it has, I don't want to know.

A day's birding you want to forget?

That's easy. In 1993 an Oriental Pratincole had been seen in Norfolk and I'd taken my son David on the twitch. We went up and saw it, and the next weekend my friend Neil asked me if I wanted to go and see it but I said I didn't have time and instead would twitch a Sardinian Warbler at Dungeness and Great Reed Warbler at Elmley – new birds for me in Britain. I didn't see either bird, and when we were at Elmley a pager went off – there was a Pacific Swift at Cley. That evening I spoke to Neil, who'd seen the Pacific Swift and also a Desert Warbler, but missed the Oriental Pratincole. And I thought, 'Why are we doing this? What's the point?' I realised that even if I'd seen those birds I still wouldn't have been happy, so I gave up twitching at that moment, and since then I've been much happier enjoying birds and wildlife in a holistic way. I think there was something missing in my life, and when I met Suzanne and had our kids, I realised that twitching had been a substitute for happiness.

If I could give you a plane ticket to anywhere for a day's birding, where would you go?

Probably the Pantanal in Brazil.

If I could produce any bird for you to have a look at it, what would it be?

Great Auk? If not, Light-mantled Sooty Albatross, when they do that wonderful display.

Favourite music?

Anything from Bach to the Beatles. Not jazz or blues.

Films?

The Right Stuff, *Spinal Tap*, *Toy Story 2*, *Kind Hearts and Coronets* – an exquisite film.

TV?

Spooks, *Homeland*, any social history series on BBC4.

Favourite two books, one of which is a bird book (not *HBW*)?

Stella Gibbons's *Cold Comfort Farm* or an anthology of English poetry. Fisher's *The Shell Bird Book*.

ALAN DAVIES AND
RUTH MILLER

Alan Davies and Ruth Miller offer guided birdwatching trips in the UK and abroad. Together they hold the record for the most birds seen, worldwide, in a single calendar year. Both were born in the 1960s.

INTERVIEWED BY KEITH BETTON

Where were you born?

Alan: Dolgarrog, in the Conwy Valley, north Wales. A great place to explore woods and streams, where my interest in birds was already growing. We moved to Conwy when I was seven, and it was even better! Our house was opposite a large field and beyond that lay Conwy Mountain and beyond that the sea. So many habitats to explore.

Ruth: I was born and grew up in a little village called Cudham near Sevenoaks in Kent. Although it's not that far from civilisation, we lived down a rough track off a country lane on the North Downs and were surrounded by woodland, farmland and open chalk downs. It was the perfect place to grow up if you loved being out and about in the countryside. I would climb over the garden fence and escape into the woods to spend all day wandering around, only returning home when I was hungry. I used to end up miles away from home, but I never admitted to my parents just how far I explored.

What was your first experience of birds?

Alan: Seeing birds on hill-walking trips in Snowdonia which I found rather boring. But my parents encouraged the bird-spotting as a way of making the hill-walking more fun, and to stop me moaning. I can still remember seeing a Red Kite in north Wales in the early 1970s when they were still rare birds – how times have changed. My first notebooks from the late 1960s show my first sightings of Redwings and Fieldfares and some more vague reports such as 'saw a hawk' and 'flock of Tufted Duck on Conwy Estuary' – almost certainly Wigeon!

Ruth: We had plenty of berry bushes and fruit trees so there was a lot of natural food in the garden. We also had several feeders and put out bird food, so I remember there were always plenty of birds to-ing and fro-ing in the garden and right outside the back door. But the birds that really caught my attention were the Bullfinches that came to feed on the apple buds in next door's orchard each spring. I could sit at our kitchen table and look up at them as they stripped the buds off the trees just over the fence. I remember wondering how a bird could possibly be such a gorgeous, almost unreal, pink/red colour, and how clear the demarcation was between the patches of colour, as if each area had been masked off with tape as they'd been coloured in. I looked up these rosy birds in my bird book and learned that they were Bullfinches. That was it; I was hooked.

Who influenced your birdwatching in those days?

Alan: My parents were interested in birds and wildlife, but were not keen birdwatchers. I was able to persuade them to take me to lots of birdy places in those early years; in hindsight, they must have been pretty tolerant! For keen hill-walkers, a trip to Suffolk would not have thrilled them, I am sure! We joined the local bird group, the Cambrian Ornithological Society, and went on some of their field trips. Here I met other birdwatchers, all a lot older than me, so no kindred spirits, sadly. However, a chap called Peter Dare, the county recorder at the time, recognised my passion for birds and asked me to join him on some of his birdwatching days out. These were often to Point Lynas on Anglesey for prolonged sea-watches when Peter would log every single bird passing, but we never saw any scarce birds! It was years later I realised our poor optics in those days were not up to the job. We saw a lot of 'unidentified skua species' and 'unidentified diver species'! In the late 1970s I met Trevor Jones, a local lad and keen birdwatcher, and he had a car, so my birding changed rapidly.

Ruth: I have my parents to thank for encouraging me to learn about the birds, butterflies and plants all around the area where I grew up. We'd usually go for a family walk at the weekends and Mum and Dad were very good at sharing their enthusiasm for and knowledge of our local wildlife. We were privileged to live on a very special area of chalk downland which is home to some rare orchids. Some of the area is now an SSSI and managed by the local Wildlife Trust. Charles Darwin lived in the next village, called Downe, and I learned in later years that he spent much of his time studying the variations in the plants here to support his theory of evolution of species. Unknowingly, I used to wander around exactly the same areas that Charles Darwin did. I don't think I will be able to make such a significant contribution to our understanding of the natural world, but I can vouch for the appeal of this quiet corner of England, even if I didn't always appreciate it quite as much as I should have at the time.

Was there ever a point where you thought of doing other things apart from birdwatching?

Alan: Not that I can remember: birds have always been my driving passion. Wherever I am, whatever I am doing, I am birding. One of the many wonderful things about birding is that wherever you are in the world, there are birds to enjoy.

Ruth: As I grew up I took up all sorts of hobbies ranging from horse-riding and sailing to drawing and playing the clarinet. I have to confess that birdwatching took a bit of a back seat with all these other activities on the go until I moved into London to get a job. Living in an urban area seemed to spark a contrary reaction in me. Now that I was surrounded by bright lights and all the attractions that a major city had to offer, I needed to get back out into the countryside on a regular basis. I needed my 'wellie-boot walk' fix, so I would spend time each weekend visiting a nature reserve or just getting back out into the country. I'd go for a walk and soak up the birds, and that would keep me going until the next weekend.

What was your first telescope?

Alan: A Hertel & Reuss draw tube, when I was seventeen years old: at the time it was wonderful! It had a zoom from 20–60× and was state-of-the-art at the time. I can still remember the thrill of opening the package and taking out the scope for the first time. It came in a brown leather case and in those days we never used a tripod! It was uncool to use a tripod, and we were hard-core birders; instead we used fence posts, stone walls, fellow birders or the car roof to balance the scope on. Mad. A tripod would have been so much better! That scope got me a lot of birds. I remember an early trip to Loch Ryan and it was only through my scope we could pick out the drake King Eider on the far side of the bay. Those with Nickel Supra scopes, the other popular make at the time, had to look through my Hertel & Reuss.

Ruth: A Swarovski 65mm telescope with a 20–60 zoom, but I had to wait until I was in my forties for it! It totally transformed my birdwatching, allowing me to get really close-up views of the birds. Once you start birding with a scope there's no looking back. It's not enough just to see the bird; you want to enjoy that detailed view every time. It can be a pain carrying it around, though; I still don't think anyone has yet invented a really good way of comfortably carrying a scope and tripod and still keeping it instantly usable. Perhaps having a personal scope-bearer is the answer!

Did you join the YOC?

Alan: Yes, at about ten years old. I still have my *Bird Life* mags. I got really excited when the latest issue of the magazine arrived, and read it from cover to cover in one go. It was great to know there were other youngsters out there who were passionate about birds. Sadly, I never met any other local members to share birds with in those early years.

Ruth: No, I never really wanted to join groups when I was young. A missed opportunity, I now realise.

Did you join the RSPB?

Alan: I moved on to the RSPB from the YOC and again loved reading *Birds* magazine. In those days the reports from the reserves were always the first thing I read. For a young birdwatcher in north Wales it was amazing to read of such exotic birds as Marsh Harriers and Avocets.

Ruth: Yes, I became a member of the RSPB in my twenties when I moved into London. It provided that vital birdwatching lifeline while I lived in the London suburbs. I'd regularly drive out to reserves in Kent and Surrey, and pored over the quarterly magazine, *Birds*, as soon as it arrived.

What about your time at school?

Alan: I was a closet birdwatcher, so most of my friends never knew! I got up early and went birding, then met friends later in the day, claiming I had only just got out of bed. A teacher, Miss Williams, at my primary school in Conwy, was very encouraging and had a keen interest in birds. All my essays in English language were always about birds, and in art I tried drawing birds, but quickly gave that up! At comprehensive school a lot of the classrooms overlooked the Conwy Estuary and I was regularly in trouble for looking out of the window.

Ruth: None of my school friends was particularly interested in birds. They all lived in towns, and I was the only one in my circle who lived in the country so, during my secondary school years, I envied them the novelty of having street lights, pavements, and – great excitement – shops right on your doorstep. Living out in the sticks as a teenager just wasn't cool. With my typical contrariness, it was only once I lived in a town myself that I realised how special my rural home was.

In those days, who were your main birding companions?

Alan: I birded on my own mostly, though in my late teens I birded a lot with Trevor Jones from nearby Llanfairfechan. As mentioned before, Trevor had a car and we were off chasing birds all over the UK! Crazy fun days, sleeping in the car and spending evenings in the pub and birding/driving all day. I have so many great memories from those early twitching days: the Belted Kingfisher in Cornwall and my first trip to Scilly were with Trevor.

Ruth: I'd still go for walks at the weekend with my parents or I just went out into my local woodlands on my own. It wasn't considered a cool thing to be interested in birds, so I'd keep my interest to myself. I'd hide my binoculars in my pocket or keep them under my coat until I was sure that no one was looking or that I was far enough away from home to not meet anyone I knew! It seems so silly now to be so self-conscious. I wear my binoculars with pride in the most unlikely birding locations now, but I guess it was all part of growing up.

Were you a twitcher at any time?

Alan: Yes. As I said before, Trevor had a car and we went off all over the UK in the late 1970s, and I also did a lot of twitching in the 1980s with Ken Croft from Holyhead, Anglesey. Ken and I clocked up a lot of great birds all over the UK and enjoyed many trips to the Isles of Scilly, where we experienced the highs and lows of the Scilly season. I think my maddest twitch was to drive to Cornwall for a Great White Egret – they were rare in those days. We heard about a Great Spotted Cuckoo in Kent, drove through the night, only to dip on it the next morning, then shot up to Humberside to score a Blue-cheeked Bee-eater before driving back to north Wales!

Ruth: Not really. I didn't know any twitchers or move in those kinds of circles. The nearest I came to it was during The Biggest Twitch around the world with Alan in 2008, although, despite its name, it wasn't really a twitch; it was more of a year-long bird race. I will twitch the occasional bird now with Alan, though don't consider myself a real twitcher; I'm lucky enough to have done a lot of birding overseas. That means I've seen many vagrants that occasionally turn up in Britain in their rightful homes. I enjoy meeting other birding friends and acquaintances who've also gone to look for a particular bird, but I don't enjoy joining a mass of people who may be elbowing each other aside in a bid to get a better view of a bird that's found itself off course in this country.

What early bird books did you have?

Alan: *The Observer's Book of Birds*, the Hamlyn guide – an awful book with misshapen birds, then Heizel, Fitter and Parslow – which was much better!

My parents had the *AA Reader's Digest Book of British Birds* and I spent hours reading this. It only showed the breeding distribution of birds, so I thought the birds could only be seen where the colour was on the map! This led to some strange early records, such as going to Scotland to tick Common Gull, only to come home and see one in the field opposite our house. Any book on birds was collected at any opportunity! Pocket money was for one thing only: saving up for the next bird book. An early edition of John Gooders's *Where to Watch Birds*, which I still have, was a real eye-opener. This did not show me what the birds looked like, but where I could see them! My parents must have hated that book, because from then on all they heard from me was, 'Can we go to…?' There was one place in this chocolate box of a book that really grabbed my young interest: an RSPB reserve called Minsmere in Suffolk. The huge list of possible birds you could see at this one reserve seemed impossible to a young birder in Wales. Finally, pester power worked and we made a pilgrimage to Minsmere and, oh boy, did it live up to the book. Lifer after lifer came thick and fast; Avocets and Marsh Harriers were no longer just pictures in a book, they were in my bins! We also saw a breeding plumaged Water Pipit – I'd never even heard of one before. My father spotted 'a Dunlin that has gone mad', as he described it. I looked over to see a phalarope spinning in circles! Our sighting was confirmed by one of the wardens as a Grey Phalarope; yet another lifer!

Ruth: My earliest bird books were Ladybird books given to me by my parents: *British Birds and their Nests*, *Garden Birds* (with those inspirational Bullfinches on the cover) and *What to Look for in Spring/Summer/Autumn/Winter*. I've still got them on my bookcase! Then I progressed on to the *AA Reader's Digest Book of British Birds*. These were great books to learn more about birds, but not exactly the sort you could take out in the field with you. I guess I only really came of age when I got my first copy of the Collins Guide. What a brilliant book! We now have hundreds of bird books all over the house. Both Alan and I love books in general and bird books in particular, so we don't need much encouragement to add more to our collection. We have all the field guides of the countries we've been to, and all the ones we've yet to visit that we take off the shelf and drool over from time to time.

Were you good at keeping notes?

Alan: No. I started so many notebooks, as early as eight years old, but usually let them fall by the wayside, only to start a new one sometime later! I love looking back at the notes that I did keep, so I really regret not keeping my notes going. While working for the RSPB in Cumbria protecting Golden

Eagles I kept detailed notes on the eagles' behaviour and saw some amazing things, including the female taking a Badger cub in broad daylight! This was the first ever documented record of this! So do keep notes. Nowadays, our notes take the form of our blog on The Biggest Twitch website (www. thebiggesttwitch.com/), so I can look back and relive my birding moments since 2008.

Ruth: No, I've never been any good at keeping notes. I'd start a new year full of enthusiasm with a new notebook for January and I'd give up halfway through the month. I'd try to pick it up again in the autumn, but would lose momentum again after a short while. I ended up with a large collection of slightly used notebooks.

When did you first go overseas birding?

Alan: In the 1980s Spain was my first trip and I headed for the Pyrenees. I remember being very disappointed as we drove for a couple of hours and the best bird we saw was a Cuckoo! Once we got off the main roads, though, we soon began to see such amazing birds as Hoopoe and White Stork. It was a brilliant week with so many new birds and I was hooked on foreign birding. For years I had said that seeing birds in the UK was the only thing that mattered, but how wrong can you be! This was quickly followed by Venezuela, which was a bit of a shock to the system, to say the least. But I'd opened a very big chocolate box and the lid did not go back on! South America is where the birds are, and I have been lucky enough to return many times.

Ruth: Working for Cadbury Schweppes in Africa allowed me to travel widely and see some great birds on my business trips. But I didn't really admit my interest in birds to my colleagues. In fact, my boss at the time, Richard Birch, was also a keen birdwatcher and he didn't admit it either. So there were the two of us with all these wonderful birds all around us and neither of us admitted to being interested in them. What a waste! We've since come out of the closet and gone birdwatching many times together in the UK. However, my first serious overseas birding trip only took place after meeting Alan Davies through our mutual work at the RSPB. We went to the Gambia first of all, as a gentle introduction for me to the wonderful birds overseas, and later we visited Namibia, which is still one of our favourite countries. Then I went with Alan to Ecuador, and having opened the Pandora's box that is South American birding, I've never looked back. However, things really took off to quite a different level with our big year of birding, called The Biggest Twitch, in 2008, when we birded in twenty-seven different countries in just twelve months!

Tell us about The Biggest Twitch. How did it come about?

Alan: Not sure! We had reached a point when we were not doing enough birding, despite working for the RSPB, and it was time for a change. Iain Campbell of Tropical Birding had a part in getting the 'break the world record' going. We had first thought of taking a year out of work to go birding, visiting lots of places and seeing lots of dream birds. We shared our idea with Iain and he said those fateful words: 'If you are going to do that, why don't you do something useful and set a new world record?' It snowballed out of control and we found ourselves doing it! But it was the best thing we ever did.

Ruth: Blame it on the Birdfair. We came back from there one year so full of ideas of where we wanted to go birding and so little time to do it. We were both really busy at work and not doing enough birding. There was only one thing for it: to give up working, sell the house, pack our bags and go travelling. Neither of us could think of a good enough reason to talk ourselves out of it, though our original plan for a gentle trip around the world escalated hugely after mentioning it to a friend of ours, Iain Campbell, who suggested we did something 'useful' like setting a new birding world record. It all rather took off from there …

As soon as we'd decided to try to set a new world record, we had to rethink our itinerary. We had to be in the right part of the world at the right time of year to maximise our bird numbers, so we zigzagged around the globe rather a lot. Our feet hardly touched the ground as we raced around the world enjoying amazing birds and accumulating a huge bird list.

What were the high and low points?

Alan: High points – seeing so many great birds day after day for a whole year; seeing stunning places; meeting amazing people; having so many experiences that we had never dreamed of; seeing a Harpy Eagle; seeing twenty-seven countries in twelve months: the list could go on and on! If I had to choose one moment, it would have to be seeing the Harpy Eagle. When I was about eight years old, I was given a book on birds of prey and in it the picture showed a Harpy eating a monkey. I was hooked! It took a long time, but finally I got to see my most wanted bird.

Low points – finishing the year of non-stop birding was a massive downer. So was having my video camera stolen in May, and having to leave so many wonderful places when we wanted to stay longer!

Ruth: There were so many wonderful moments, but for me the highest of the high points has got to be when we broke the world record after just ten months on the road. Bird number 3,663 was a Bluebonnet parrot which we saw as it enjoyed the watered greens on a golf course in Leeton, Australia on 31 October 2008. We finished the year with a total of 4,341 species recorded in the course of that year. It was such an incredible experience and it is still hard to believe even now that we really did it.

There really weren't many low points; we were incredibly lucky with good health and trouble-free travel throughout the year. However, perhaps the lowest point was just as the clock struck midnight on 31 December and the most amazing year you could ever imagine came to an end. We celebrated our record-breaking achievement, but that meant that our year-long quest was now over and we had to return to Britain. It was January, the country was in the throes of a recession, and we had no money and no job. It wasn't a great moment. But even if we couldn't spend more time travelling and birding, we could do the next best thing – writing our book about it. So that kept us busy until we set up our own birdwatching tour company, called The Biggest Twitch Ltd, of course!

How much did it cost you?

Alan: A lot more than we budgeted! We owe a huge debt in many ways to Ruth's mum for helping us finish the adventure by financing the last part of the trip.

Ruth: About twice as much as we'd expected! We're better at birding than we are at budgeting, and in August when we came back to the UK briefly for the Birdfair, we checked our finances and realised that we didn't have enough money left to finish the year as we'd originally planned it. So, we slashed our plans dramatically, reducing the number of flights we would take and cutting out many of the countries we'd hoped to visit. Instead we focused on countries which still had plenty of new birds but where we could stay longer and more cheaply – India, for example. Even this still cost more than all the money we had left, so I had the awkward task of asking my mother if the 'family bank' could possibly bail us out and enable us to finish our year. Luckily, Mum appreciated that we'd put absolutely everything we had into our record-breaking attempt and that, if we didn't finish, we'd have spent it all and achieved nothing. So she took a deep breath and got out her cheque book. So, thanks Mum, we couldn't have done it without you!

Was it all sweetness and light on your year of The Biggest Twitch, or were there times when you felt like killing the other one?

Alan: Oh yes! Usually at non-birding times. Travel brings out the worst in me: being tired, not seeing birds and being herded about with lots of other people puts me in a bad mood for sure! In Canada, I can remember us falling out badly, but we had just been robbed and were not finding many new birds – it seemed it was all going wrong. But that soon passed, thankfully.

Ruth: Yes, there were a few times when I could have cheerfully pushed Alan off a cliff. But if you spend that much time in one person's company, you can expect a few fireworks. But we really didn't argue that much as we travelled on The Biggest Twitch because we were both so focused on seeing the birds and helping each other to see the same birds. It was only when we came home at the end of our year and sat down to write the book that we started to argue. We couldn't agree on how to write a single sentence together, so that's why we wrote separate chapters in separate rooms – but that probably made it a better book, as well making sure we stayed friends.

If you could do it again, what would you do differently?

Alan: Have a bigger budget. We would spend more time in Asia and less time in Europe. We would pace ourselves at the start of the year, January 2008 was just mad and exhausting but we did see over one thousand birds in just one month. We would spend more time in each country so spend less time flying and more time birding. Columbia and Uganda would be on the list, as would Thailand, Malaysia and Borneo. We would see even more birds! Our target would be 5,000 species, but that would be very hard to reach. In 2008 we had no major health problems or travel disruption – either or both could have knocked us off target. Would we ever be so lucky again? Also, flying costs have risen dramatically since 2008, making it even tougher.

Ruth: Second time around, we'd probably arrange our itinerary differently. Ideally, we'd spend longer in Asia, which would give us high numbers of birds – this was a region that got cut out the most in 2008 because of the number of flights needed. We'd probably spend less time in Europe, to keep the costs down. In 2008, we deliberately visited quite a few European countries as we know the birding areas well and as we know people like to hear and read about places they have experienced themselves. Oh, and we'd pack fewer unnecessary spare clothes too! It's amazing how lightly you can travel when you put your mind to it.

Would you do it again?

Alan: Oh, yes! Without a moment's hesitation! If there are any potential sponsors out there for The Biggest Twitch Two, do please get in touch, we would love to have another crack at it.

Ruth: Oh, yes! We might have to rob a bank or win the lottery first, though, or perhaps persuade a TV channel that we'd make the perfect subject for a series. If there are any sponsors out there, do give us a call!

Are there any other challenges you'd like to do?

Alan: Visit more countries that I have not yet visited and see a lot more of the world's birds. I would like to set some twenty-four-hour bird race records: the UK one has stood for a long time and it would be fun to try in other countries and perhaps set a new world record for a twenty-four-hour bird race. Again, sponsors, don't be shy, do drop me a line!

Ruth: We have lots of ideas for more road trips and more countries that we'd like to visit, and it would be fantastic to revisit some of our Biggest Twitch locations, though perhaps with a bit more time to enjoy them. I would love to visit Antarctica and also Kamchatka, to name just two, so hopefully one day we'll get there. As former RSPB staff, both Alan and I really appreciated visiting a number of ecotourism projects around the world, and we have given talks about our experiences. We were very impressed with what could be achieved in various countries, often without much money but always with a huge amount of dedication and drive. It would really be rewarding to make a point of visiting as many conservation initiatives around the world as we can, to help raise their profile and celebrate their achievements with as wide an audience as possible.

Where will you be in ten years' time?

Alan: Still birding hard! Running a very successful bird tour company, The Biggest Twitch, and having seen a lot more birds. Still holding the world year list record, but perhaps with now over 5,000?

Ruth: I'd like to be still enjoying travelling around the world and looking at birds in ten, twenty, maybe even thirty years' time as long as I'm still fit and healthy. In ten years' time, I'd like to think we'll still be sharing wonderful birding moments with other people through our successful bird tour company. And I would love to have written and published several more books. I've got more of my birding/walking/cake-eating guides called *Birds, Boots and Butties* in the pipeline, which provides the perfect excuse to get out birding and try out new cafés and cake, all in the name of research, of course. And they say everyone has a novel in them, so I'd like to find mine and publish it. Who knows? It could be a bestseller!

If I could introduce you to several people you've never met – alive or dead – who would they be?

Alan: Jesus – it would be great to get the story from the man himself and interesting to get his take on how the world views him. Also Eric Cantona – one of my all-time football heroes and a very interesting man.

Ruth: Queen Elizabeth I – a strong female leader in what was very much a man's world in Tudor times; Winston Churchill – I would like to meet the man who was such an inspiring leader in the war years; William Shakespeare – I'm fascinated by language and we owe so much of our current English language to him; Ronnie Barker – a brilliant comedy talent; the scripts he wrote are simply timeless.

What about within the realms of wildlife and natural history?

Alan: Ted Parker – the American birder who sadly died young. Ted sounds like he was a driven birder, and it would be great fun to go birding with him. Jim Clements, who held the world year list record before us. Gilbert White, the naturalist, whose book *Natural History of Selbourne* I read as a child.

Ruth: Charles Darwin – as he was a near neighbour, only separated by a hundred years or so; Alfred Wallace – to keep the balance in our discussions about the evolution of the species; Roger Tory Peterson – his concept for bird guides transformed the way we identify birds in the field today, and I'd love to know more about where he got his inspiration; David Attenborough – most people of my generation grew up with his iconic TV wildlife programmes, and *Life of Birds* was a particular favourite; Phoebe Snetsinger – I would love to meet the lady in person to discuss her passion for listing.

Where is the best place you've ever been birdwatching?

Alan: A very tough one. My best country would be Ecuador, for the sheer variety of birds and habitats. My favourite place would be the Pantanal in Brazil, which is teeming with birds and wildlife.

Ruth: I fell in love with the Pantanal in Brazil – my favourite place out of all the locations we visited on The Biggest Twitch. It's teeming with exciting wildlife, which allows you to get quite close to it so you can enjoy really wonderful views. A few of my favourite species include Hyacinth Macaws, Jabiru Storks and Great Potoo. I'd go back there in a heartbeat.

Where is the worst place you've ever been birdwatching?

Alan: A rubbish dump in India to see Greater Adjutant Storks: the stink, and the horrific sight of poor people trying to eke out a living from the foul place … but we got the bird!

Ruth: KwaZulu-Natal in South Africa. That may sound a bit harsh. In fact, we thought it was a great place until one day we found ourselves caught out in a bushfire. The strong winds whipped up the flames until they completely surrounded the building in which we'd sought shelter from the fire. All we could see on every side was raging flames licking at the windows; the air filled with smoke and it was hard to breathe. We really thought we were going to die. We survived, but fourteen unfortunate people were killed that day in the same bushfire, so we realise we had a very lucky escape.

If you could go birding to one more place in your life that you've never been to, where would that be?

Alan: Colombia, with so many birds and so many new lifers in one country! South America is mind-blowing for a birder and I would love to get to Colombia soon.

Ruth: That's tough to answer, as there are so many amazing places I've yet to see, but New Zealand would be right up there. It was one of the countries on our original itinerary for The Biggest Twitch that we had to drop when we cut back our plans. It has some very special birds, though not enough new ones to justify the cost to visit New Zealand back in 2008. So we've unfinished business there …

What is your favourite bird group?

Alan: Tough, but eagles have a very special place in my heart.

Ruth: My favourite bird group would be the Ramphastidae family of toucans, araçaris and toucanets. The Toco Toucan is the mascot of The Biggest Twitch and we even took a toy Toco Toucan around the world with us in 2008. It's a distinctive, iconic family, and individual birds are always fascinating to watch.

What is your most wanted bird?

Alan: Steller's Sea Eagle. Having seen Harpy Eagle this is now my most wanted; an iconic eagle with a huge beak, that is found in wild places!

Ruth: My most wanted bird would be the Kiwi, though I don't mind which species I see first. When I was little, my father used to clean my school shoes every night with Kiwi polish and I always wanted to see the funny-looking bird on the tin. I'm not sure if it was supposed to represent a particular Kiwi species, so I'd be very happy to see any Kiwi at first hand.

On a desert island, which piece of music would you take with you?

Alan: 'Atmosphere' by Joy Division; I can listen to it over and over again.

Ruth: It would have to be the Hebrides Overture by Mendelssohn. I studied this for music O-Level, when we managed to persuade the teacher that we needed to visit Staffa in order to pass the exam! Our ruse worked, and I've loved visiting the Hebrides ever since.

If I'm allowed pop music, I'd probably ask for 'Private Investigations' by Dire Straits. It's a long track with plenty going on, so I could kick back and relax as I listened to it in the shade of my palm tree.

Favourite film?

Alan: *Top Gun*: a good fun adventure with a great soundtrack. I always wanted to be a fighter pilot – not really!

Ruth: It would have to be Monty Python's *Life of Brian*. I've watched that so many times I can recite half the lines, but it still makes me laugh until I cry.

Favourite TV show?

Alan: *Fawlty Towers* never fails to make me laugh out loud.

Ruth: I don't watch much TV, but could I have a boxed set of *Have I Got News for You*? on my island? It's always very funny, even if you catch an old programme, and provides a great window on what was going on in the news and on the political scene at the time.

Favourite non-bird book?

Alan: I've not read many! *Touching the Void* – the true climbing adventure story was a gripping page-turner. At an early age, Henry Williamson's *Tarka the Otter* was a big hit, but there are quite a few birds in there so perhaps it doesn't count?

Ruth: It would have to be Tolkien's *Lord of the Rings*. I do love a thick book you can really immerse yourself in. I've watched the film, but nothing compares with reading the book and imagining the scenes and characters for yourself based on Tolkien's words.

Favourite bird book?

Alan: *Kingbird Highway* by Kenn Kauffman – inspiring stuff. It's the tale of Kenn giving up school and going in search of the USA year list record by hitch-hiking across the States living on a dollar a day. It's a good training manual for The Biggest Twitch, though we didn't (knowingly) eat cat food on our big year.

Ruth: No question – *The Big Year* by Mark Obmascik – it was the perfect training manual for The Biggest Twitch. And the book is so much better than the film. You really get the idea of three men driven to see more birds. If you've not yet read it, rush out and buy a copy and forget anything you saw on film.

REBECCA NASON

*Rebecca Nason is a freelance ecologist, tour leader and photographer based on
Shetland. She was born in the 1970s.*

INTERVIEWED BY KEITH BETTON

What were your first binoculars?

As far as I can remember, they were Opticrons. I later moved on to Swarovskis,
which I loved, and had until 2004 when I got my treasured Leica Ultravids,
which I still use today.

What were your first bird books?

My first memories of coveting a bird book – *The Shell Bird Book* which I loved
and used until my early teens. Our house was full of bird identification books
for all over the world.

Tell us about your early birding history.

At primary school, which was a short walk from our house, I won an art
competition with a drawing of a Kingfisher which became the school emblem
and was printed on all the sport T-shirts. I remember picking a Kingfisher
because I had been elated at finding out what was taking all the goldfish from
our garden pond. At the age of eight, I set my alarm for dawn and over several
mornings, peering, on tiptoe, from the bathroom window, I spotted the
culprit: a flash of electric blue and a Kingfisher appeared at the pond side. I
knew they lived on the local river, but to see one in my own garden was so
special.

I had a couple of amazing pets when I was at primary school. The first was
a family of Grey Squirrels. After hearing the screeching of brakes outside the
house one day I went to the road to see the horrific sight of a flattened adult
female Grey Squirrel and a tiny baby, alive, hovering over her. I scooped the
young one up and, on hearing alarm calls, looked up to see three more young
Grey Squirrel heads poking out from an old Jackdaw nest in the tree opposite
our house. Dad got a ladder and placed the young one back in and we left

them. By the end of the next day, the calling of the young animals had become unbearable. Clearly no adult was in attendance so I was allowed to collect the four young ones up and bring them home before they starved. I put them in a box with cotton wool and placed them next to the Aga in the kitchen. I fed them warm milk with a pipette, and bread, as I tried to save them. One died, but the other tiny babies continued to grow. Soon we had three playful young squirrels we could play with and cuddle. They had to stay in a cage in the house, and we would often take them to climb trees in the garden for exercise. I took them into school once and let them out in the classroom. One ran behind a cupboard and took an hour to coax out with nuts. After about six months we released them back into their dray and, as far as we know, they survived and thrived. At the time I just wanted to save their little lives: Grey or Red, it didn't matter.

My second pet was the much-loved Michael, the Starling. I was walking around the side of the house one day and came across a number of small naked, blind chicks on the gravel. The Starling nest in the eaves of the house had collapsed and the contents had been thrown out in the cold. The young were so vulnerable, and I was desperate to try and save them. Again they went into my cotton wool ball-lined box in the kitchen and were force-fed by pipette. As expected, as they were only a day or so old, they all died, bar one: Michael. I've no idea why we called him Michael, but it seemed to fit him well. Soon his eyes opened and he started to show signs of feathers. Later he flew freely around the house and, seeing me as his mother, would follow me to school in the morning and sit on the school fence. He was great. We took him to Wales with us on our family holiday. I remember he sat looking out of the back window of the car during the journey, and it was in Wales he learned to catch his own food, running and grabbing spiders as they ran over rocks. Unfortunately, as expected I suppose, life ended tragically for Michael. He had no fear; I'd often have to get Dad to go and retrieve him from the shoulder of an innocent passer-by or a guy having a pint outside the local pub. One day, he followed us to the local stables where my sister and I were learning to ride. I spotted a ginger cat streaking across the yard with a small bird in his mouth – it was Michael. My mum ran after the cat, which dropped him. He wasn't dead, and Mum had to finish him off behind the stables. Michael was about one – but what a year he had!

How did you get into birds?

I have always been into birds, having grown up in a family of birdwatchers and naturalists. My grandparents on my dad's side were very keen birders and went all over the world on annual trips with Ornitholidays in their later years together, though my granddad died when I was eighteen and my lovely Gran lived for another twenty years. Dad and his sister grew up as keen naturalists, enthused by their parents, and as a result my sister and I grew up as equally enthusiastic naturalists with a passion for the outdoors, travel and nature. My dad is the birdwatcher and is keen, but barely gets to lift his bins these days as he has a very successful, but very full-time, business.

We would always try and go on one family holiday abroad a year when I was young and one walking trip to Wales or the Lake District at Easter. Our family holidays were the event of the year and full of excitement and great family experiences, all revolving around birds. Although my mum and sister, Tolly, were also keen naturalists in my youth, it was my dad and I who were keen birdwatchers and had a more serious interest. Mum and Tolly used to lose interest quickly, they said partly as a result of not having as good eyesight as us and not being able to pick up birds in the field easily.

I used to be called 'well-spotted Becca' on holidays, as I would always be calling birds out from the car and identifying birds immediately that I had only viewed briefly in bird guide plates. I seemed to have a natural ability to spot and identify birds, something that I'm sure is partly inherited. Learning about and progressing with birds came naturally; it's been one of the few areas I have not had to make a deliberate effort to learn about. I seemed to soak it all up subconsciously when I was young. I'd be up early and keen to see as many bird species as possible wherever we were.

I also became keen to photograph birds too, borrowing Dad's Nikon F4 and 70–300mm zoom lens and stalking birds on holiday at any given opportunity. I got into the finals of the Young Wildlife Photographer of the Year twice in my youth, which inspired me further. The photographs were of a female Palestine Sunbird taken in Israel in 1989 and a Green Heron taken in Florida in 1988. Of course, this was a long time before digital photography.

Our holidays when I was young were never relaxing times, when we had time to chill on sunbeds in the sun; instead we were always trying to fit in as much as possible and see as much of the country as possible and as much wildlife and as many birds as possible. I used to complain sometimes that we were in need of a holiday once we had returned to the UK after our holidays, but of course, I'd not really have done too much differently, with hindsight.

Our adventures went a bit far sometimes, though. I recall when we came close to having our plane shot down as we approached Kabul, Afghanistan, while on our way to India at half-term one year. Dad had bought us all return tickets to Delhi off Ceefax, a place we often used to look at for good flight deals abroad in the 1980s. We flew via Prague, and joined an Afghan Air flight. After an hour or so, we started to descend and looked out to see pine trees stretching as far as the eye could see. Soon we landed in Moscow – which was not on our itinerary! Our 'Afghan Air' plane had no hostesses, and a load of drunk Russians boarded in Moscow. I wasn't happy by this point, and the smell of aviation fuel was nauseating. Mum was sitting with my sister behind me in between two fat, drunk Russians, and Dad's seat didn't have a seatbelt. A few hours later we were over the mountains heading down towards Kabul. We could see plumes of smoke puffing up from the bare red hills. All of a sudden our aircraft was flanked by two helicopters and flares were fired from them, diverting potential missiles from hitting our aircraft! We circled the airport like an eagle from 30,000 feet before landing on the tarmac and were walked across the airstrip to a shack of an airport under armed guard. All this before we flew on to Delhi and our 'holiday'. I remember asking why we couldn't go to Portugal like everyone else. However, the trip was amazing. We spent most of the week at Bharatpur and made it back to school in Cambridgeshire for first thing Monday morning. My week had been vastly different from my peers' that half-term break.

As a young girl, certain family trips abroad really stood out – The Gambia, Iceland, Trinidad, Guyana and Chile – but I remember raving about a holiday to Scotland in 1984. That trip was wonderful, camping and driving about the most beautiful parts of Scotland. I turned ten just before that summer holiday and joined the YOC at Loch Garten after seeing my first British Osprey on the nest. I collected postcards of Scottish birds along the way – Curlew, Whimbrel, Red-throated Diver – and remember sitting in the back of the car looking out at the heather-clad terrain and dreaming of living in Scotland and marrying a Scot. Although I am going to marry a Mancunian, not a Scot, it seems like fate that I have recently moved to live on the Shetland Islands after the influences from that summer.

Were you a lone girl birder?

I did not have like-minded peers in my youth, which is a shame. Perhaps I should have gone on field trips with the YOC, but instead I was more of a loner with my interests and enjoyed spending time with my family and my parents' friends. They seemed much more interesting.

Although my family life was great, life at school was quite harsh between the age of twelve and fourteen – those often difficult years. I hated school. I was at a local comprehensive. I was the youngest in my class, perhaps a little more immature than my peers, and I certainly had a much more interesting and exciting time outside school than most. As a result of my extracurricular interests, some youthful weight gain, my quiet nature and my interesting life at home and abroad, jealousy-based bullying kicked in. Bullying was not dealt with well at school in those days, and the teachers struggled to handle it (maybe that hasn't changed). I suffered from relatively mild mental and physical bullying by boys, but mainly by a few girls, for a couple of years, during which time I became more and more unhappy and saw my academic life spiralling downwards. Although I remember actually becoming stronger as I grew up with it, moving to a different school at the age of fourteen to start my GSCEs was a great move for me.

My life changed from being bullied and failing academically to being popular and seeing the academic benefits that come from being a happier child. I'm so pleased my parents made that choice to move me when they did. I was a day scholar at a boarding school in Essex. I loved it so much that I occasionally boarded myself, and I often stayed well after the classrooms closed for the day. I stayed for four years, doing my A-Levels there too, and went out with the Head Boy during school, my gap year after school and through my first year at university in Winchester.

So my youth was two extremes: perhaps experiencing both being so low and being so high has made me the strong person I am today. I do remember smacking the two boy bullies the day I left that old school, which gave me an immense sense of satisfaction! As for the girls who bullied me, I feel sorry for them, as I did at the time. During my early teens my birdwatching largely fell by the wayside, apart from those annual family adventures.

What did you do in your gap year after school?

I went to Crete that summer on a girls' holiday. It was a very different holiday than I was used to; we did lots of sunbathing and all came back very brown. The best part of the holiday was getting the girls to walk Samariá Gorge on my eighteenth birthday.

I worked part-time in [ladies fashion retailer] Monsoon in Cambridge to raise funds and the next spring my boyfriend and I went to Vancouver and west coast Canada for two months, camping, hitching, birding and volunteering for a conservation organisation that was trying to save a pristine area of Vancouver Island from being clear-felled. We camped in the stunning temperate rainforest where we woke to the sounds of Common Loon and daily bear sightings, and worked hard to help build a wooden trail through the forest to showcase the area to the public and make them aware of what was about to be lost. Amazingly, Clayoquot Sound was later to be saved from the axe and became a protected reserve. We met some great like-minded fellow travellers. I decided Vancouver was the one city in the world I would love to live in if I had to leave the UK.

However, I also had my most frightening wildlife experiences to date there. We had several very close Black Bear encounters and I learned what it felt like to be scared encountering a potential predator. I slept with a knife (Dad's Persuader fruit-cutting knife) under my pillow that trip.

Were you academic at school?

Far from it. I had little time for school, and when I was there I often wasn't paying attention in class. It was not until I had decided to stay on at school to do A-Levels that I began to actually want to be there and make an effort to learn and do well. Then I became more academic and started to achieve in areas I thought impossible previously. I could never decide what to do at school (follow science or art), so I kept both as much as I could, desperate not to be channelled too young and close opportunities. I ended up starting a degree in art, geography and environmental studies, later dropping the art to concentrate on the science. Being an ecologist and wildlife photographer now gives me satisfaction in both art and science, which I love.

What about university?

I had such a good time at university, in one of the best cities in Britain, Winchester. From my hall of residence I looked out over the destruction of Tywford Down as bulldozers moved in to destroy a chalk downland hillside awash with butterflies and orchids, and a place I'd spent many happy times as a child. The destruction of that stunning SSSI, and witnessing the ripping apart of that hillside for a motorway outside my bedroom window, definitely influenced me politically and environmentally. On the odd weekend before the destruction had fully taken course, I went with a couple of friends to see the protestors, who had built shelters and were living on the hillside. I gave the activists biscuits and newspapers and wished them well. I think of that beautiful hill, and the raping of the land, every time I drive along the Winchester bypass.

Winchester was full of great pubs and I spent my student time burning the candle at both ends, working hard to achieve a good degree while being out most nights on the town enjoying myself. Somehow I managed both pretty well. The icing on the cake was walking down the aisle of Winchester Cathedral to receive my upper second honours degree.

I feel a great affinity with Hampshire and the south coast, aside from having gone to university there. My birding grandparents lived in Chandler's Ford, Dad was brought up in Southampton, and we spent a lot of weekends over the years visiting grandparents and birding, walking, looking for fungi in the New Forest or fossil hunting along the coastal cliffs.

At university, geography field trips were highlights. I went to Poland and Ireland as well as several Field Studies Council (FSC) field centres on trips. I remember skinny-dipping off the west coast of Wales with a big group of friends at 2am one October after leaving the local pub, and doing a project on Rough Periwinkles. The most influential trip was to Skomer in September 1995. I was totally smitten before we even landed on the island, and even more so after finding a fluffy young Manx Shearwater among the bracken. I remember thinking at the time that that day-trip meant more to me than to anyone else in my class, many of whom seemed unfazed by the visit. For me it was my future and when I left university the next year I was accepted as the first long-term volunteer on Skomer Island, living and working there for three incredible months. That cemented my desire to work within conservation, birds and, where possible, small UK islands.

Did you have other hobbies apart from birding?

I spent hours looking for fossils in the Barton Beds along the Hampshire coast throughout my youth. I had a very keen eye for fossils. I remember finding my dream fossil, the rarest one illustrated in my Barton Beds fossil book, *Notorhynchus primigenius* (a type of shark tooth with many serrated edges, looking more like a jaw than one tooth) from 40 million years ago (relatively young as far as fossils go). I also found a whole 40-million-year-old pine cone washed out onto the shore at low tide after a storm – a really rare find and one which is now displayed in the Sedgwick Museum of Earth Sciences in Cambridge. Just talking about fossil hunting makes me want to head for the south coast and start hunting – I've not been for years. The elation in finding a good one is on a par with finding a good bird.

When not birding for work, pleasure or tour leading, my other hobbies I guess include being a licensed bird ringer (for the BTO), moth trapper and a real 'foodie' with a fondness of good food and wine, fuelled by foreign travel. I love Asian and French cuisine in particular. I have also inherited a strong appreciation of art and certain antiques. My Gran was an antique dealer with her own shop in Romsey, Hampshire. I collect antique bird prints. My favourite possession aside from my bins and photo gear is a seventeenth century Bellarmine jug.

When did you start twitching?

I have never been a hard-core twitcher, but I certainly am a twitcher and happy to be so. I love natural history in its many forms; seeing a rare vagrant to Britain is exciting and uplifting. I find twitches tend to be happy (I don't dip too often) and social and educational occasions and you get a great sense of place, visiting at short notice a part of the UK you might otherwise never have visited.

I find people who knock twitchers, for their passion and hobby, sad and irritating.

I think the first birds I twitched were Bluethroat and Wryneck along the north Norfolk coast and a flock of Waxwing when I was at primary school. I did not start really going for rare birds specifically until 1997. After doing a MSc in Conservation Management in Suffolk I met more serious birders, had a birder boyfriend and started twitching more frequently from then.

My best year was the year I first visited the Scillies, 1999. I spent the summer volunteering on the islands carrying out seabird census work and seabird ringing with Peter Robinson, ex-chief investigator of the RSPB. I loved the Scillies and all the seabird and boat work. I have stepped foot on most of the islands on the Scillies and stayed overnight camping on Round Island and several other uninhabited islands. I returned to the Scillies alone that autumn and camped alone on the Garrison for two weeks in October. I saw the Blue Rock Thrush, White's Thrush and Yellow-billed Cuckoo among many others and I also found the great appeal in trying to photograph rarities. I was befriended by the autumn rare bird photographers and shared evening curries and many rarity tales, culminating in me later investing in a decent lens and a tripod. That autumn rekindled my passion for bird photography which has stayed with me ever since.

So you became a Scilly autumn regular?

I spent a large part of 2000 in Cape May, USA, birding, and returned to the Scillies in autumn 2001 after some seasonal ecological contracts. I was free from a failed relationship and as keen as ever on birding on Scillies and bird photography with a great experience of USA birding under my belt.

I spent the autumns of 2001 and 2002 on Scillies photographing rarities, and selling my pics at the log which was very enjoyable. I imagined I'd return to the Scillies every autumn, however I took a job as assistant warden and Seabird Officer on Fair Isle in spring 2003 and my life changed again and took a different path.

In 2003 I left Fair Isle to go to the Scillies before the end of the season on Fair Isle. This was to cost me dearly as I missed the Savannah Sparrow and

Siberian Rubythroat and saw very little on Scillies that year. I made sure I stayed on Fair Isle the next year until the end of the season; the highlight was, of course, the two birds new for the Western Palearctic, Rufous-tailed Robin and Chestnut-eared Bunting. I was one of just a handful to see these birds and got the best images so I was totally made up that autumn. I found the birding on Fair Isle and Shetland had the edge, for me, over Scillies and I changed course north.

What were your Fair Isle highlights when you worked there?

I hadn't been to Shetland or Fair Isle until March 2003 when I flew up for the new contract. I had no idea at the time how taking that job would change my life and how I'd fall in love with the place and my future husband that year.

I've spent every autumn ever since along with my partner, Phil, a firefighter, birder and bird ringer I met on Fair Isle in autumn 2003, at least in part on Fair Isle and Shetland. I have had a lot of luck birding on Fair Isle since, returning annually to visit too, so long may that continue. I have not given up the Scillies; we have been on August pelagics off St Marys a few times and plan to go in the autumn as a change from Shetland soon.

When working on Fair Isle I had a birder's dream job. I censused a third of the Island on rotation with the warden and other assistant warden every day during spring and autumn. As well as recording daily common migrants, you inevitably also found your own rarities in such a hot-spot.

In the summer we scaled and descended 200ft cliffs, or headed out in a Zodiac, to monitor and ring seabirds. The job was very physically demanding, I was so fit during those two years and the birding rewards were high; those were special years.

Your best finds to date?

In 2003 I found a Yellow-breasted Bunting and Citrine Wagtail during the same day on Fair Isle. In 2008 I found a White's Thrush. Other finds include Olive-backed Pipit, Hornemann's Arctic Redpoll, Blyth's Pipit, Booted Warbler, Radde's Warbler and co-finding Pallas's Grasshopper Warbler. On Shetland, Phil and I re-found the Pine Grosbeak last February which was just incredible and I found a nice Black-throated Thrush last November near Lerwick, while out counting Greylags.

What is your current British List?

427, not that big. I pick and choose what I go for and have missed many an easy tick. My last tick was the Cape May Warbler on Unst, Shetland, in October

2013. I imagine a life on Shetland, and regular Fair Isle trips, will see a few more notched up before too long though, and I might have to twitch the mainland occasionally and when visiting family in the south.

Where did your career take you post Fair Isle?

Post Fair Isle was always going to be a difficult move, what can beat that job? However I decided to go freelance and became an Ecologist and Bird Photographer as well as Wildlife Tour leader for the Travelling Naturalist in 2005. A mix of these jobs has been immensely rewarding, varied and meant I have managed to combine my passions of ecology and conservation, birding, bird photography and travel going all at the same time! I knew that after birding Fair Isle, locating somewhere else was always going to be so diluted in comparison. I love the Suffolk coast and we lived there for a few years, however we were drawn back to the Isles and decided we would move to Shetland full-time.

I have had hundreds of images published in books, journals and magazines and had numerous cover shots. I am represented by several picture agencies, am an Associate of the Royal Photographic Society and came second in last year's Scottish Nature Photography Awards in the 'Natural Abstract' category. I am one of the top ten photographic contributors in *Birds of Scotland* and many of my images appear in the famous New Naturalist book series. I am a full member of the Chartered Institute of Ecology and Environmental Management and carry out numerous bird surveys across the UK, including offshore seabird work.

I have led trips to Lesvos, Greece for eight years now and also lead trips to Spitsbergen as well as closer to home with Shetland-based wildlife guiding and photo-tours.

So you have recently moved to Shetland?

Yes. It was a big move. I arrived last October after finishing an ecological contract in Essex, and Phil came up earlier in July to start work. We were looking to buy somewhere on Shetland over the past few years and found our dream house in Lerwick in 2011. We had to wait to sort work out before actually moving here two years later. We had been looking for a more rural location, but the Georgian sheriff's house overlooking the harbour and Bressay is more than we ever hoped for. When I first visited the house pre-purchase I watched a pod of Pilot Whales from the bedroom window and knew this was the house for us. It is a good size, has excellent views and a garden with lots of mature vegetation in it as well as being big enough for a few 60-foot nets for bird ringing! We did not want to wait twenty years for vegetation to grow up to a reasonable size, so are delighted to have a ready-made bird-welcoming garden. The garden in central Lerwick has so much potential and has already had three Hawfinch in the Hawthorn tree together, an OBP (Olive-backed Pipit) on the lawn, a Citrine Wagtail overhead, a few Yellow-browed Warblers, Barred Warblers, Waxwing and Ring Ouzel – and this is just the beginning! We have a respectable house list of ninety and counting, with breeding birds such as Whimbrel, Red-throated Diver, Great Skua and Arctic Tern as standard. Tundra Bean Goose was the last 'house tick'.

Your birding passions at work and play have been in a largely male-dominated domain. How have you found this?

Fine, on the whole. I'm a very social person and have always got along very well with men. It's also easy to get on with people you have so much in common with, a shared passion. There are few women birders, of course, which is a shame. There are a lot of women working for conservation organisations and in research but most are birdwatchers, not what I would class as serious birders. I hope that changes. I'd like to see many more women taking up such a pleasurable pastime; I'm sure there would be a lot of happy male birders as a result too! I certainly stuck out when birding on the Scillies as well as Fair Isle, particularly in my twenties, but I was treated with respect and the same way as everyone else, and I can't say I didn't enjoy the attention either.

I have only had one really worrying incident that was a direct result of being female. It happened in 1997 when I went back to Poland with a female friend from university. We had loved the field trip to Poland so much we vowed to return again to explore more of the country. We started in Krakow which we had been to before and loved, and then we got a train across to the east and a village near Białowieża Forest. My friend wasn't a birder but I was keen to try and see some Polish forest birds and persuaded her to join me. We paid for a guide to take us into the forest early one morning with the hope of seeing what was at the time a lifer for me, a Black Woodpecker. The guide knew of a nest and all looked rosy. However, things took a rather sinister turn in the forest, with our guide getting carried away at guiding two young female English birders. He started really letting himself down. It was basically quite mild sexual harassment, with an ever-present flicking of a large Swiss army knife. He was full of bravado that he knew the forest and we hadn't a clue which way was out or how to get out. It made for a very unpleasant morning's birding.

What is your favourite music?

I have quite varied taste in music. I spend hours singing away to the radio on my long journeys for work when south, often arriving in my destination with a sore throat! I'm currently enjoying Ellie Goulding, the Arctic Monkeys, The Killers, Green Day, Mumford & Sons, some 1970s romantic pop and 1990s Brit pop. I also love some good jazz over a pint or two.

TV?

I'm enjoying the *Shetland* series by Ann Cleeves at the moment – Jimmy Perez walked past my house in the last episode. I love *Game of Thrones*; that hits a lot of what I like in an 'epic' plot. When we eventually get Sky installed on Shetland maybe I'll be able to enjoy Season Four. I also enjoy *Coast*, *Masterchef*, *Hotel Inspector*, *Antiques Roadshow*, *Deadliest Catch* and *24*.

ROBERT GILLMOR

Robert Gillmor is an artist. He was born in the 1930s.

INTERVIEWED BY MARK AVERY

Would you call yourself a birder?

I think I would have done years ago. Today, I don't think I really quite fit the term. I birdwatch all the time but don't cross the road to see a rarity. Of course, I'm enormously interested in birds, and I get just as much pleasure from seeing birds on those feeders there [pointing out of the window] as I do going down the road and looking for some poor rarity that's gone off course. I do watch birds keenly, though. I suppose I am still a birder, really, but I am not as serious an ornithologist as I was years ago.

We did a lot of work on bird behaviour when travel wasn't possible, and twitching hadn't been invented. One was really more interested in what birds did rather than ticking them off. I've never been a ticker or a lister – that doesn't appeal to me at all. I don't have the faintest idea what my British list is, or find it interesting.

Can you remember your first pair of binoculars?

I had a monocular which used to belong to my grandfather and I had one of those long brass extending telescopes too. And later I had a shore shooter's tripod so that I could lie down and look through it. I can remember lying down and watching Pink-footed Geese in Iceland through my long brass telescope and drawing them.

Nowadays I have a pair of Swarovskis. I had an old Zeiss pair for years and years and years. The new binoculars are just so much in advance of anything I've had before.

What was your first bird book?

A book that meant something to me was *Our Bird Book* by Sidney Rogerson, illustrated by Charles Tunnicliffe. I first saw it in the flat of my best friend's aunt. The plates by Tunnicliffe were wonderful and this book was fantastically expensive. It was two guineas – quite beyond me in 1947! I was just enthralled by the illustrations. But I would have had *The Observer's Book of Birds*. I saw this wonderful book and must have pleaded with parents, or an indulgent grandmother more likely, and eventually I got my hands on a copy. I was always very keen on Tunnicliffe – it was obvious he knew the birds at first hand.

How young were you when you got interested in birds?

I was interested in birds right from the start. My grandfather, Allen W. Seaby, painted the plates for Kirkman and Jourdain's book, *British Birds* (published in 1935), although he wasn't simply a bird artist, they were a very fine set of plates and were quite influential in their time. I used to spend many hours in his studio watching him work. He had retired before I was born, but I was fortunate to be able to spend time with him and see him at work. He was a printmaker doing colour woodcuts using traditional Japanese methods, which he had, with one or two others, more or less pioneered at the end of the nineteenth century. He carried on with those throughout his life, as well as painting.

In his old age he was very deaf, and we used to go walking in the countryside round his house and I was his ears for bird songs and calls.

I would just spend time at one end of the studio while he worked at the other, and then as I got a bit older, when I was ten or eleven, he introduced me to, and showed me how to do, things like oil painting. Not that I ever did much oil painting, but I did my first one or two pictures with him. He was a very fine art teacher all his life. He had been at Reading University, first as a student and then was co-opted onto the staff and eventually ended up as Professor of Fine Art there. All that was long before – I just knew him as an old deaf man working away in his studio at prints and watercolours. He never did much bird painting after the Kirkman and Jourdain series, except towards the end of his life. He died in 1953. In the early 1950s, Brian Vesey-Fitzgerald was commissioned to write some books on birds for the Ladybird series. My grandfather was asked to paint the pictures for these and they were some of the first bird paintings he had done for a long time. Anyway, the first book was pretty successful and he was asked to do a second. At that point I was at school at Leighton Park in Reading, where we had a very active ornithological group. We went on a trip to Skokholm – Peter Condor was then the warden – and I did a lot of sketching, including of Puffins flying around the cliffs.

How old were you?

I suppose I was about fifteen. Anyway, I came back with my sketchbook and my grandfather was having trouble with a painting of a Puffin in flight for the next Ladybird book. So with the aid of my sketchbook, and his skills, we produced this painting of the Puffin in flight which was fine. He was going to do a third series but by that time he was in his mid-eighties, I suppose, and was taken ill and never completed the job. It was then that the publisher realised that they had been dealing with a man in his mid-eighties. He had the most wonderful handwriting and all their contact was done by correspondence – I doubt there was ever even a phone call – and they were very surprised to discover his age. Roland Green did the third volume.

Grandfather just about saw my first published illustration in *British Birds*, which was in 1952. That was a plate of shearwaters for a paper by Max Nicholson on shearwaters in the English Channel. All this came about because at Leighton Park there was, on the staff, a master, J. Duncan Wood – not now a name known to anybody. He was a very fine field ornithologist, and he had been at the school as a pupil when they had gone to Skokholm at the time when Ronald Lockley was setting up the bird observatory there. They were involved in helping to build traps, like the Heligoland trap there, that sort of thing, and then he went off to serve in the Friends' Ambulance Unit in China. He came back to teach at Leighton Park and ran the Bird Group which met on Friday nights. He was an absolutely wonderful person to be taught by. The Bird Group was a very strong group in the school and a whole number of pupils went on to become quite significant ornithologists on the British scene. You would have worked with James Cadbury ...

Indeed. He was at Leighton Park?

Yes, he and I were at school at exactly the same time. Humphrey Dobinson and a group from the school were largely responsible for setting up the bird observatory at Cape Clear in southern Ireland. We went out there several times. Another was Stephen Baillie, now at the BTO.

Anyway, the point was that, at that time (in the early 1950s) Bernard Tucker, who was the editor of *British Birds*, was not a fit man and Duncan was asked to take over from him for a period. This was incredible for us, as we were having regular Bird Group meetings and ringing, and going off on expeditions with someone who at the time was right at the centre of ornithology. We had a complete set of *British Birds* in the school library, right back to Volume 1, and used to spend a lot of time looking at those. We would see proofs of the photos and articles in advance of them coming out. And he gave us copies of the rather tatty proofs of photographs – I still have them in my files. I never throw anything away!

When Max did his paper on shearwaters he wanted illustrations of several species. It's surprising, when you look back at that period, that although today there are bird illustrators around every corner, there weren't then. You had the leading lights such as Peter Scott, Donald Watson, Keith Shackleton and Charles Tunnicliffe, but these were the professionals – Richard Richardson was here in Cley later on – but there was nobody you could instantly ring up and ask, 'Can you do a line drawing of this species?'

So anyway Max wanted a drawing and I did a couple of versions of it and got it right and it was published. That, of course, was an enormous fillip to me.

Also through the Bird Group we were getting to know all sorts of people because we had visiting speakers on Friday nights. The Bird Group meetings ran on very formal lines. We had a Secretary and a committee. The Secretary wrote up the minutes of each Friday meeting and these were read, approved and signed by the Chairman at the following week's meeting. We had to introduce a speaker and someone had to thank the speaker, so we were getting the most fantastic experience for those of us who would later go into organisations, committees, lecturing, all that sort of thing. We were terribly fortunate.

I was Secretary for a while, and Chairman later on, when I was in the sixth form. It really was a remarkable time. Duncan left to go to Geneva, to the Friends' Centre, while I was at school, so for the last two years we were pretty much running the Bird Group ourselves but the headmaster, who was a very keen botanist, got involved, so we were very fortunate there.

All this time, I was also involved with the Reading Ornithological Club (ROC) which had started in 1947. Duncan Wood and Eric Watson, who was the brother of Donald Watson and at the Botany Department at Reading University, and one or two others started the ROC which they named by mixing up the names of the Oxford Ornithological Society and the Cambridge Bird Club. The meetings were all held at the university and still are.

I was terribly keen as a small boy on birding, and I was taken to ROC meetings when I was about twelve, even before I had started at the senior school in 1949. I was far too young to be a full member, but in the minutes of the meeting in January 1948 there is a note that says that Robert Gillmor, on account of his extreme youth, be encouraged to come along as a visitor. At the ripe old age of thirteen I was elected as the Club's first junior member and within ten years I was Honorary Secretary. I carried on being a member of the Club until we moved to live in Norfolk in the late 1990s. I was

Secretary and Chairman and then eventually President and now an Honorary President.

The first two ROC annual reports were just duplicated on a Gestetner, as so many things were in those days, but for the third one they thought they ought to have some sort of cover. By then I was at Leighton Park discovering linocuts and a linocut of mine, of a Canada Goose, was used for the first real report of the ROC for 1949, when I was just thirteen years old. And I was, as you can imagine, thrilled and have been doing covers ever since!

I particularly remember a talk by James Fisher where I asked the great man to autograph a paperback copy of one of his books. Some sixteen years later he opened the inaugural exhibition of the Society of Wildlife Artists (SWLA).

Our art master, Harry Stevens, was also a great influence and help.

I had an extremely fortunate start in the bird world and was remarkably lucky to be around at that time.

While I was still at school I got to know Ken Simmons and Derek Goodwin. Ken arrived on the scene in Reading, after he had done National Service abroad, at the time when Little Ringed Plovers had come to the Reading area. We were very protective of them because we knew that they were pretty scarce, and here was this chap wandering around, looking a bit suspicious, obviously showing an interest in them too. Eventually I got to know him and we worked together and, of course, I illustrated a lot of his early papers on bird behaviour. So I was looking at, and sketching, Little Ringed Plovers and Great Crested Grebes, with more than just a casual interest.

We used to go to Virginia Water, at weekends, to Derek Goodwin's house where he had aviaries. These were two amateur behaviourists, in the sense that they weren't in university departments, but both were highly regarded in the bird behaviour world. Derek's papers on bird behaviour of crows and pigeons, Ken's on Little Ringed Plovers and grebes, were very highly thought of.

I actually illustrated a paper on Magpie behaviour for Derek in *British Birds*. It was my second lot of illustrations for *British Birds*, and would have been in the early 1950s. I did these illustrations for Derek and it's always appealed to me that ever after that he did all his own illustrations – and very well too. For he really understood bird behaviour, and the extraordinary positions that birds get into, and while I saw some of them and worked from his sketches and notes, I think I did him a good turn there, getting him to do his own from then on. But with the shearwaters for Max, and the Magpies for Derek, and some work for Ken, I was, as a schoolboy, seeing my work in print, which was, of course, an enormous encouragement.

Because of the ringing and everything I got involved with the BTO – which at the time was based in a tiny office in Oxford. I did a lot of work with them because, again, there weren't that many people they could get hold of. The same happened with the RSPB. I started working with the RSPB when they were at a tiny office too, in London.

In Eccleston Place near Victoria?

Yes. You see, the way that happened was because the ROC was one of the first clubs in the country to hold an annual film show with the RSPB and, at that time, it was Frank Hamilton who was working in London and he took the film show around with a screen and a projector. We had these shows in the Great Hall of the university. I got to know Frank quite well. And then the RSPB might, like the BTO, want a little bit of illustrating and so they'd ask me. I'd do it at once and send it off. So I started doing drawings for the RSPB, line drawings, of this and that, very early on.

Were you good at school?

No, I was pretty hopeless! Well, I was good at art and I got the necessary number of A-Levels, but I didn't have a distinguished academic career, at all.

What subjects did you do in the sixth form?

Oh, heavens, I don't know – apart from art! It must have been geography or something like that.

I went from Leighton Park to university in 1954. But before going to university I went on a two-man expedition to central Iceland on a travel scholarship for two months, following in the footsteps of Scott and Fisher – but that's a whole other story. And then came straight back to the Reading University Fine Art department, where I was for five years.

I went back to Iceland in 1956, to watch the geese coming through northern Iceland from Greenland, and then in 1957 we formed the university Exploration Society and mounted a major four-man expedition to Spitzbergen, during which I made a 16mm film and took photographs. After that I started to do lectures all over the country.

When I was at university, having a wonderful time, I was, by then, practically running the ROC and doing a lot of work with the BTO and the RSPB. I also illustrated my first book – David Snow's *A Study of Blackbirds*.

I had joined the BOU as soon as I could. You couldn't join until you were eighteen, and you had to be proposed and seconded, and Ken and Derek did that. So I've been a member of the BOU ever since I was eighteen. It's a long time – terrible, isn't it?

I met David Snow, who wasn't much older than me, at a BOU dinner when we were sitting at the same table. He had been doing his D.Phil. on Blackbirds in the Oxford Botanic Gardens and one thing led to another and he asked me whether I'd like to illustrate his book. Well, I didn't let on at the university art department that I was illustrating a book, I don't think it would have gone down terribly well. Once it was published I took it along to the Professor and he was a bit surprised, really, but he took it quite well!

That, inevitably, led to other things. But of course I had to decide what I was going to do. At that time, going into teaching was a very obvious thing to do. My course was four years, and then I stayed on and did a fifth year of art teacher training. I spent some time as a trainee teacher at a school in Wokingham and they more or less offered me a job but at the same time I got a call from my old headmaster at Leighton Park telling me that the art master, who had taken over from Harry Stevens, was leaving to join the family laundry business. He knew I was about to come on the market and he asked me round for a chat. He knew me from being an ex-pupil. At the end of our talk in his study he offered me the job of coming back and running the art and craft department. A school like Leighton Park always likes to have Old Leightonians on the staff. So I had to decide one way or the other, and it wasn't a very difficult decision. And so to my astonishment, five years after leaving, I found myself in the staff room at Leighton Park with all the teachers who had tried to teach me their subjects a few years earlier.

But in that five years a whole new art and craft department had been built. So, in 1959, I had a large new art studio, a big woodwork shop, a big metalwork shop, a pottery, and book-binding and printing departments. All these were run by individuals who came in from outside as part-time staff, and I was in charge of all this, which was pretty incredible, but I sort of coped.

Of course, the headmaster was only too happy to hand over the Bird Club to me. And this was when we started going down to Slimbridge to take part in the Wildfowl Trust (now the Wildfowl and Wetlands Trust) school competition, which tested knowledge of waterfowl. We trained very hard for those competitions and there would be a coach full of us heading to Slimbridge for the competition in March each year.

That's why Leighton Park won it so often! Even more often than Bristol Grammar School did. I remember taking part in it when I was at school there in the early 1970s. It was on identification of European ducks, geese and swans.

As a boy I had first visited Slimbridge with my buddy Nick Blurton-Jones – he had joined the Wildfowl Trust in the year it formed in 1946, I joined the following year. And we went down and stayed for a week at a time in the winter holidays on a narrowboat on the canal. The first time we turned up, they were expecting two large ornithologists from Reading and two schoolboys in shorts arrived – they were a bit amazed.

What did you do there? Count geese?

We were actually studying triumph behaviour in geese where the male impresses the female by attacking another male and then making a lot of noise about it …

In the captive collections?

Yes, that's right. I remember making a short 9.5mm film of Nene geese doing triumph displays.

Hugh Boyd and Frank Mckinney were both staying on the narrowboat at the time. Because the Wildfowl Trust was a very small outfit, we got to know everybody down there, including Peter Scott, and years later, when I was back teaching at Leighton Park, I was able to ask him to contribute to the exhibition, in 1960, that led to the founding of the Society of Wildlife Artists. I was absolutely enthusiastic about bird art and its history and rather frustrated that it was extremely difficult to see original work by bird artists other than the odd card or book illustrations. The artists were scattered around the country and never got together and I thought, 'Why can't we join up and have an exhibition?' Having done those illustrations for Max Nicholson, he invited me to go as his guest to the Christmas dinner of the British Ornithologists' Club at which the Club 'entertained the bird artists', which I thought was fantastic. And at that meeting were a dozen artists, like Keith Shackleton, David Reid Henry and Eric Ennion. Peter Scott was in the chair.

Dear old Max – he started such a lot of things, didn't he? In this case, he didn't start the SWLA but he provided me with the opportunity and encouraged me in all sorts of ways.

I already knew Peter Scott, and I'd got to know Donald Watson and Richard Richardson, and so I wrote to them and said, 'Why don't we all get together and have an exhibition?' I knew the Reading Art Gallery people and, using the ROC, I approached them. They responded enthusiastically. Eric Ennion was very keen and so we planned an exhibition by contemporary bird painters.

The first exhibition was in the autumn of 1960 and, through Peter, it was opened by Lord Alanbrooke who of course was a great naturalist as well as being a war hero. We had this wonderful opening afternoon in Reading and it was a jolly good exhibition and was taken on tour for a year by the Art Exhibitions Bureau of the Federation of British Artists (FBA). It was highly successful, so they extended it for another year.

By that time we were well in with the FBA and Eric Ennion and I had started talking about founding a society. Eric and I had a great time planning the SWLA. I particularly remember that, at a BOU conference in Chester Zoo, we cut some of the lectures and sat in the sunshine making notes and planning how we were going to do it, and writing constitutions and picking founder members. Maurice Bradshaw from the FBA was very keen, and so all went ahead and in 1964 we had the inaugural exhibition of the Society of Wildlife Artists, opened by James Fisher.

But it was surprising, you see, how few bird artists there were. For the 1960 exhibition, we had about thirty-five people, but in 1964 there weren't many more. There aren't, sadly, many survivors from that time; myself, Ian Wallace, John Busby and a few others.

And all that happened when you were teaching at Leighton Park?

Yes, while I was teaching I was also doing my own work because I thought it was good for the boys to see that I was a serious artist as well as trying to teach them. And I was also doing an increasing amount of outside work, illustrating for the RSPB and the BTO, book illustrating and all sorts of things. It came to the point where I had to decide which way it was going to go. I was either going to stay in teaching or stop and become a freelance.

At the time I was living at home with a studio shed in the garden, and there seemed to be enough work to keep me going as there really weren't many illustrators around in those days. So it wasn't such a difficult decision.

I went and talked to the headmaster and said that I thought really the time had come for me to push off. And he was brilliant and said that he could understand that but that the art department was growing and the school needed another member of staff to run the craft side with me doing the art side, so he asked me to stay another year and run art four days a week and show the new guy the ropes. So that's what we did. They got a splendid new chap in to run the craft side and it worked out very well. I eased into freelancing in that way. I retired in 1965 and I've been messing around ever since. I was extremely fortunate.

Do you have a favourite bird?

Not so much a favourite bird, but a favourite type of bird. What really appeals to me, and it's very much to do with my approach to art these days, is birds with bold, simple plumage patterns. Particularly long-legged wading types like Avocets, herons and Oystercatchers, because they are so suited to what I am really interested in as far as picture-making goes. Herons have always appealed to me, Avocets once I got to know them. They just give me a great deal of artistic pleasure and, of course, they are the ones that I largely represent in my print-making. Fussy little speckly brown jobs that live in the middle of bushes are useless!

What's your favourite place to go birding?

Nowadays north Norfolk is my favourite place.

Sue and I got married in 1974 and holidayed here as a family for eighteen years – a stone's throw, literally, over there. If I lobbed a rock over your car it would land on the roof of the cottage we used to stay in here in Cley. When the children fledged, we thought we'd like to find somewhere up here, but never thought we'd find anywhere in Cley, but fortunately we were looking just before the housing boom and we found this place, having got to know this area through our holidays. Earlier, I used to come and stay with my great friend Noel Cusa at Letheringsett.

I went to Titchwell in 2010, the year before the RSPB centenary, because I was commissioned to do a painting of an Avocet family for one of the groups who wanted to produce a print to sell for fundraising. And I would be out in the Parrinder Hide by about eight o'clock and spent a week drawing and sketching. So I really fell in love with the area. When I first came, Richard Richardson was still active and he would always come over to Noel Cusa's, where I would stay, and we would have a great evening talking about bird art and drawing. Noel had a good collection of pictures in his sitting room and, as he was a great friend of Tunnicliffe's, had a number of his originals. We always enjoyed that.

If you could do anywhere in the world to see a bird or paint, where would you go?

Although I did do quite a bit of travel in my early days, to the Arctic, USA and East Africa, I absolutely hate travel these days. I never travel more than five yards if I can help it. I loved the Arctic each time I went. The Antarctic would be rather fine – all those penguins. I illustrated two books on penguins and I had to make do with those I could find in this country to sketch. It's surprising how many penguins there are in captivity, what with Bourton-on-the-Water and London Zoo.

How many individual pieces of bird art have you produced?

I haven't the faintest idea! There would be several thousand drawings altogether and then all the paintings. All the calendars – I did calendars for over thirty years and latterly there were two calendars a year, needing twenty-four paintings, so those add up. Between 2010 and 2011 I was doing designs for postage stamps for the Royal Mail. In 2010 I had done twenty-four stamps, five book jackets and a few other things which added up to thirty-two pieces of work, which is over one a fortnight. Eleven days for a piece of artwork doesn't leave you much time for anything else!

I don't know, and most people who aren't artists don't know, how you go about producing a piece of art. So let's take the stamps. The Royal Mail, presumably, told you which birds they wanted?

Yes, they gave me the list of birds, and the first couple of sets of stamps were based on the commonest species in the RSPB Big Garden Birdwatch. Later there was a set of waterbirds. I would have liked to include a heron – because I've always liked drawing herons – but they wanted a Greylag Goose, so that's what I had to do.

And presumably they gave you a deadline?

Oh yes, they were very strict about deadlines.

And how long before the deadline did you finish?

To start with, there were six stamps and I suppose I was approached at the end of 2009. They issued them in September 2010 so I was working on them right through the start of the year and so I had about six months to do them.

I knew the format – the size and shape of the stamp and where the Queen's head and caption were going. I knew that the bird had to take up quite a lot of the space and I had to think of it 'stamp-sized' as well as the size I was going to paint it, which was about A4. I'd want to show off the characteristics of the bird, and it would need to be quite bold because it was going to be reduced so much. And then I would just draw it …

Easy for you to say …

I would sit down with a pad of blank paper and start doodling. They only wanted one individual bird per stamp. So, I would draw a Blue Tit and put a nice big oak leaf near it and send off the drawing. I had very little trouble with them at all. It was all quite straightforward.

In the second series of six stamps, they wanted a Magpie. And I said you can't have one Magpie – one for sorrow – and so they said all right, paint two! The Magpies are so boldly marked birds it wasn't too difficult to make two birds stand out.

Were you doing lots of other work around the same time?

Well, I have a commission to do the dust jackets for the Collins New Naturalist series and there are three of them a year, so I would have had to fit them in too. I love doing the book jackets for Collins, and every one is different, so I have to make myself an instant expert on subjects about which I know absolutely nothing. I started in 1985 and I've done over fifty. And there would have been other regular projects I'd have to fit in. I did manage to do one print of my own that year, but that was the only one.

You clearly differentiate in your head between the things you are commissioned to do and the things that you want to do as an artist …

That's absolutely right, and well spotted! When we moved here I had pretty much decided there would be no more book illustrating – I felt I'd done enough, I'd done it for forty years and I wanted to get back to print-making. I'd stopped active print-making in the mid-1970s because when I got married in 1974, and we had a family, I had to make a living and print-making was a bit of a luxury. I had to spend all my time doing commercial work – at those times it was quite tight. By the time I moved here I had a few blocks cut which I could use to print editions and get the press set up. I hadn't used the press for years. It was built in 1860. I determined that I was going to start print-making again. I pretty much stuck to my determination not to illustrate books, but I was asked by two friends to whom I could not say no. I illustrated Mike Shrubb's book, *The Lapwing*, and Duncan Wood wrote a biography of H.G. Alexander, which I illustrated with black-and-white drawings along with Ian Wallace – but those are the only two books I have illustrated since moving to Norfolk. I had illustrated a book for H.G. Alexander in 1974. I did do drawings for the Norfolk Bird Report for many years until they changed editors, and they now seem to have lost my address.

So since we moved here I have really got back into print-making in a big way.

What is it with linocuts that appeals so much to you?

Well, what I like is that it makes me simplify so much, and to see the essence of whatever it is I am doing. I just want to get down to the absolute basics of the bird, which is why I like birds with simple, bold plumage patterns. As a medium for something like a book jacket, it is the bold clarity of the design which is good, so that a book jacket can hold its own against the myriad other books on the bookshelves.

I like taking a group of Avocets or Oystercatchers and arranging them to create their own particular pattern or design. I'm not trying to do a field guide illustration at all – it's something much simpler and bolder. I never seemed able to do that in a painting because in a painting my work always got rather traditional and detailed, and while it was all right it never satisfied me in quite the same way as doing a bold three-colour print.

And I rather enjoy the technical side of linocutting, particularly playing around with the colours. You see, if you use three colours, and you print one colour on top of another to make an extra colour, there are all sorts of things you can do. If you count the paper as a colour and have three coloured inks to play with, A, B and C, then A+B, A+C and B+C, and A+B+C and the paper adds up to eight colours. And that's rather fun. And I think it's just more me, frankly!

How do you decide what your next piece of art will be?

I don't know, really. Well, a lot of them arise from things I've seen out in the field which stick in the mind as subjects. Or things that I've sketched, so I look through sketchbooks and bring several together in a composition.

Would you call yourself a bird artist? Or a wildlife artist? Or simply an artist?

Inevitably I am seen as a bird artist. But I have done other things.

Yes, I know you have quite a liking for Hares.

Yes. And Badgers …

The Avocets of the mammal world? And you have shown me stamps that you have designed based on domestic livestock – sheep, pigs and cattle. Are there other aspects of your work that we birdy people may not have seen that would surprise us?

Quite frankly, I would never have done a set of linocuts of pigs, sheep and cattle if I hadn't been asked. But when I came to do them I discovered a whole new field of subjects. The Badger Face Welsh Mountain Sheep is wonderful! And one I had never heard of – I didn't know it existed! It was great fun.

I did a jacket for a book entirely devoted to the Pike, and another book on fishing. I also did a book jacket featuring a village cricket match once.

I was very lucky, I thought, that in my office in The Lodge I had 'Two Turtle Doves' by you above my desk.

Yes, I can tell you what that was about. The RSPB produced four sets of prints and I did the spring ones. I think there was a Song Thrush, Swallows, I can't remember the third one, and those Turtle Doves. And that was done – I looked it up – in 1972.

I must say, it gave me a great deal of pleasure having your work on my office wall – and sometimes in a difficult phone conversation I could look up and see those two peaceful doves on the wall. Is there a time of day when you work best?

Not really. Sue always says she works best in the morning, but I don't think it makes a lot of difference to me. In the past I used to work quite a lot in the evenings.

Would you say bird art is thriving?

Absolutely: I think it is! There are a huge number of wildlife artists these days compared with fifty years ago and many are doing extremely well. The number of bird books being produced hasn't slackened in any way. There are the endless books on identification of birds in every part of the world and all of them need illustrations. There are lots of extremely talented bird artists and I imagine that the competition is very fierce. I wouldn't want to be starting in these days.

And I never really wanted to get into the field guide aspect of illustrations, but I did a bit. I was the art editor of *Birds of the Western Palearctic* for thirty years. I did comparatively few plates myself, but it meant organising everyone else.

What do you think about photography, because that is clearly taking a major part of what used to be the space occupied by artwork?

I am absolutely staggered by the quality of the images – it's just fantastic. What I find fascinating in the whole area of bird photography and bird art is that I think that now photography has liberated the artists to be more artistic and to go into areas where photography can't go. It gives the bird artist the chance to be much more original, imaginative and impressionistic, and to produce pictures with feeling and interest.

There will always be a place for the bird illustrator because it's very difficult to find a single photograph that can do the job that a bird illustrator can do for a species. A photograph records one individual at one moment of time. Very often in a position that no bird artist would ever draw, because you don't see it unless you are taking a photograph at a very high speed. And often an entirely acceptable photograph looks quite ridiculous when drawn because we don't see birds as the camera does, so the angles of the wings or something may look wrong. You can always tell if a painter has slavishly copied a photograph – at least, I think I can.

Actually, I think some of the great contemporary and innovative bird photographers may have been influenced by bird artists!

I love photography and love looking at photographs. A bird artist can learn a lot from photographs. And the finer details of plumage can often be got from a photograph when they are difficult to get in the field or even from a skin.

I hate working from skins. I've had to do it in my time, but breathing life into a skin, unless you really know the bird to start with, is jolly difficult. The great bird illustrators of the nineteenth century, John Gould, for example, had to work a lot from skins. They did their best and were remarkably successful, considering.

John James Audubon, of course, worked from freshly dead specimens, rather than skins, and that's a bit different. I love Audubon – he had a great sense of design.

I bet you wish that the cover of *Birds* magazine (now *Nature's Home* magazine) still had artwork, not photographs?

Well, that period when Nick Hammond was editor was wonderful. He gave many of us a wonderful chance. Think of Mike Warren's cover of the Pink-footed Geese, which was sensational at the time. John Busby's covers, John Paige's covers – it gave us an opportunity to produce artwork that wasn't completely traditional. Because *Birds* wasn't fighting for space on the WH Smith shelf, the title was quite tiny, so we had the whole cover to play with. It was a great opportunity for many artists to be seen by huge numbers of people. I was very lucky; I did a dozen covers in all.

So, do you wish the covers were still artwork or not, because you've avoided the question?

Well, I take my hat off to *Birding World*, which always had artwork on the covers – I think that's wonderful. When *Birds* started it went from *Bird Notes*, with a Tunnicliffe on the cover, to *Birds* size. The RSPB Council agreed that 50% of the covers would be artwork and 50% would be photographs. There was an Eric Hosking photograph and then a Talbot Kelly photograph but quite rapidly the artwork took over for about twelve years.

I think it was a shame – well, actually, I don't know. Perhaps for the bulk of the members, maybe a fantastic photograph was what they appreciated most. It would be nice, I think, to go back to a mixture. I'm sure the change was agonised over.

But I do take my hat off to *Birding World* for having artwork on the cover, and of course *British Birds* used to have artwork, but the thing it did which had a huge impact was the Bird Illustrator of the Year Award. That had a huge influence on young bird artists. I was a judge for that for many years. I think it helped raise the standard for black-and-white illustration, and of course it was all black and white back then.

Favourite artists?

In my own field of print-making I absolutely adore the work of Edward Bawden. I love the impressionist paintings, and I was terribly keen on Mary Fenton paintings.

There were two wildlife artists I got to know in my grandfather's studio. Neither is as well-known these days as they should be! The first was the Swede, Bruno Liljefors. He did the most amazing wildlife subjects. But his paintings were just damn good, and the fact that he happened to be painting wildlife but as well as other subjects. He had a fantastic eye for composition and visual memory.

Another with a remarkable visual memory was Joseph Crawhall. Grandfather had prints of Liljefors's work and a little book with black-and-white illustrations and I used to spend hours looking at that.

Another artist he introduced me to was an English artist, J.A. Shepherd, who was an animal caricaturist for *The Strand* magazine (later *Punch*), but he also had a serious side to his work. Just before the Great War he started a series of books called Bodley Head Natural History, but only the first two were ever published, because the war put an end to the projected series. For the time, his drawings were completely original and were drawn from life. He did a lot of sketching in the zoo, and produced a wonderful series of, not exactly cartoons, but witty drawings called 'Zig Zags at the Zoo' in 1894.

When I showed his work to Eric Ennion, who hadn't known of it before, he was absolutely knocked out by it – as was John Busby. They were as thrilled as I was, because here was someone who was drawing birds with a real eye for their character. He illustrated a lot of books and hardly anyone knows of his work these days – it's rather a shame.

Wildlife art has a very bad name in the wider world of art. It doesn't have a terribly high reputation – its tradition is not seen as a particularly good one, even though there are really good artists, such as Thorburn and Tunnicliffe. It has often been easier to get a painting of a dead grouse into the Royal Academy exhibition than one of a live grouse. Wildlife art has not been seen as artistic enough, but some of the younger SWLA members are frightfully good artists – they just happen to use wildlife as their subjects.

The life of a bird artist has suited you well.

Yes, very well: I have been extremely fortunate. I've met lots of wonderful people and been to some marvellous places. And I'm not sure what else I could have done!

LAST THOUGHTS

When we conceived this project, we had no aim other than to chat to some interesting birding people and to see what came of it. We certainly did not plan to collect important information which we would analyse, and from which we would draw conclusions. However, as we carried out the interviews – and, we hope, as you have read them – they stimulated some thoughts, and we have jotted ours down here.

The power of birds

What is it about birds that can enthuse such different people, so deeply and in so many ways? Would it be possible to bring together a similar book on mycologists? Maybe it would, but we doubt it – but perhaps that is because we are two of the twenty contributors ourselves!

It's often been remarked that birds experience the world in a somewhat similar way to our own species. Sound and sight are very important to birds, which is why they tend either to have fantastic plumages, or fantastic songs, or both. Most mammals – our own species is a bit of an exception – experience the world much more through touch and smell. Most of them are nocturnal, whereas birds share the daylight hours with us. This means that you can more easily identify most birds, and watch them and enjoy them.

Birds are also with us throughout the year, and are found, albeit in varying numbers, in all habitats on Earth. Whether you are sedentary or well-travelled, and wherever you travel, there is a bird to see or to hear. Watching birds can be done everywhere – watching birds in your garden, on the way to work, on business trips abroad and wherever your holidays might take you. There is no closed season, and no barren land, for the birdwatcher.

On the shortest visit, perhaps on business, to a foreign city, a birder is likely to look up and see a gull (which gull?) or a swift (which swift?) flying over, and the trees outside the hotel window, or in a local park, will have a few birds in them. They may be the familiar House Sparrows of home, or they may be completely unfamiliar species that send the birder thumbing through the pages of the local field guide to work them out.

In the UK, the number of birds that one might see in a year, even with quite a lot of effort, is manageable. The identification of 200–400 species is a challenge, but for quite a lot of people, an entertaining and perfectly achievable challenge. If there were many fewer, the task might be a bit dull, and if there were many more, it might be a bit daunting (for some of us, at least).

And those 10,500 or so species of birds worldwide make a tempting target for some. Two of our interviewees, Alan Davies and Ruth Miller, hold the record for the most birds seen in a calendar year – about 40% of all the world's species. The 5,000-species year must be entirely feasible with luck and finance. And in a lifetime it is possible to see at least three-quarters of all those species (as has Keith Betton, and he's still going strong). The world record is Tom Gullick's 9,000+ species. We estimate that the twenty Britons interviewed here have seen between 8,000 and 9,000 species of the world's birds.

No two walks in the British countryside can be quite the same for a naturalist, and this applies strongly to birds. If you visited the same place (imagine a place you know well) at the same time of day, on the same date, in the same weather, for twenty years you would be unlikely to record the same bird list in any two of those years – even if the population levels of the relevant species did not change. Even though you might record a Dunnock or Dunnocks on most visits, there would be some days when you returned home thinking, 'I didn't see a Dunnock today – how odd.'

Each visit to a nature reserve, or glance into your garden, is a sample of the wildlife available. We cannot control what we see – only be alert and try hard to spot what is there. That can bring birdwatching close to a field sport where the birder pits his or her wits and skill 'against' the birds to try to record them.

Although we all know, more or less, what we are likely to see when we go out birding, there is always that element of unpredictability about it – and that is part of the attraction. Perhaps birds, being sufficiently numerous in species, being mobile and being migrants, deliver that unpredictability in a more pleasing way than other taxa.

Twitching

Among this group of birders, some do twitch, some have twitched but no longer do, and some have never really indulged.

Although these twenty people are all birders of a sort, they are not birders of exactly the same sort.

The connectedness of the world

There is a theory that everyone in the world is connected to everyone else by a maximum of six degrees of separation. Our twenty interviewees probably know, personally, a high proportion of British birders and people involved in British birding. And if you allow friends of friends, we imagine it is a very large proportion.

The interviews established some other connections. Here are a few examples:

- Ian Newton's prized first bird book, Kirkman and Jourdain's *British Birds*, was illustrated by Robert Gillmor's grandfather, Alan Seaby.
- Ian Newton, Chris Packham and Mark Avery all gave talks at EGI student conferences that were important in determining their future careers.
- Andy Clements, Mark Cocker and Ian Newton all did their early birding in Derbyshire.
- Many of our interviewees cited *The Observer's Book of Birds* and the Peterson, Mountfort and Hollom field guide as important to them in their early years, and that was what we expected, but we were surprised by how many also mentioned the *AA Reader's Digest Book of British Birds* – Debbie Pain, Mike Clarke, Mark Avery, Lee Evans, Keith Betton, Alan Davies, Ruth Miller and Stephen Moss. That book clearly had a big impact on many of us, even though we ended up having quite widely different interests and roles in birding.
- It's not at all surprising that north Norfolk, Fair Isle and the Scillies come up often in conversation but, for one reason or another, the New Forest is often mentioned in this book too.
- Colin Tubbs was mentioned by Chris Packham, Mike Clarke and Steph Tyler as being someone who encouraged them.
- Of those of our twenty interviewees who went to university, Cambridge was the most frequented, by Steph Tyler, Ian Wallace, Stephen Moss and Mark Avery. Will it be such a high proportion in a couple of decades' time? We suspect not.
- The Spoon-billed Sandpiper is clearly a birders' bird – three of our interviewees mentioned it with great enthusiasm: Mike Clarke, Debbie Pain and Roger Riddington.

We wonder how many, and how long ago, and where, did the largest number of our interviewees assemble in the same place? Was it at a BTO conference in the 1970s, perhaps? Or on the East Bank at Cley one autumn day?

If we all had dinner together, which other linkages might emerge?

What would non-birders think of us?

We don't really care, do we? We are enthusiasts about birds and think that anyone who isn't interested in birds is a bit odd. What does it matter if they think we are odd in being interested?

However, some might look at the lengths of time and expense expended by some birders and think that they were a bit extreme in their interest. Some might even question to what extent some birders are really interested in birds at all – and whether, with a slightly different upbringing, it might have been stamps or train-spotting that captured their interest.

We rather hope that when our interviewees interact with 'normal' people some of their enthusiasm for birds rubs off, to lead to a new devotee of birdwatching.

Women in birding

Only four of the twenty interviewees are female. This merely reinforces the view of many male birders that there aren't enough women around. There were many other women who we might have interviewed (as there were men), but not many others who we thought we definitely *should* have interviewed. Perhaps you, the reader, will think differently.

Although there are women leading many of our NGOs (e.g. the National Trust, the Wildlife Trusts, Plantlife, Wildlife and Countryside Link, the Marine Conservation Society), this isn't currently true of our largest bird NGOs – the RSPB, the BTO, the WWT. And *British Birds*, *Birdwatch*, *Birdwatching* and the late lamented *Birding World* did not have women editors. Is there a woman birding journalist? Science seems a more gender-balanced profession, and that is reflected in our interviewees.

This is partly a factor of age. In twenty, and probably in ten, years' time, the list of interviewees for a book like this will almost certainly have more women in it – it may well have female authors too.

Early influences

All of our interviewees started to be interested in birds at an early age – usually younger than ten years. That was quite a striking commonality. There is no one in this list who came to birding in their middle age or later. Such people undoubtedly exist, but it may be that they are the exception. It may also be that birds have a particular appeal to young people because they are obvious, and easily noticeable, and you don't need much more than a pair of cheap binoculars and a bird guide to get started. It is more difficult for children to explore the world of mammals, fish or insects than birds – birds are easy!

Most, but not all, of our interviewees could name an important early influence on them, sometimes a parent, sometimes a teacher, who had encouraged an interest in birds. We think this was probably very important and we wonder how many brilliant contributors have been lost to the world of birding because individuals just didn't get enough encouragement from an adult early on.

The YOC (now Wildlife Explorers) seems to have played a reasonably important part in many lives, but maybe not an absolutely crucial one. And for most of our interviewees we are looking back to a time when the YOC would have been one of the few sources of information and comradeship for young birders.

Some of our interviewees had a group of friends with whom they went birdwatching as children – but not all of them. This seems to have been important for many people and, again, makes us wonder how many great contributors to birding may have been lost through the lack of a friend with whom to share the interest.

These interviews have reinforced our view that support in the early years is important in encouraging a deep and lifelong interest in the natural world. We wonder whether our generation, our society and our bird-related organisations are doing enough to encourage the next generation of birders.

The changing world

It is difficult for the authors and readers of this book to put themselves in the position of imagining what it was like for Phil Hollom to become interested in birds between the two World Wars. The British countryside was richer in many bird species (no other interviewee would have had much chance of seeing a Red-backed Shrike while doing a school examination) and poorer in others. Phil Hollom's lifetime saw at least twenty-eight species of bird become extinct on Earth. The Laughing Owl of New Zealand was the first of these, in 1914. Many other species, such as the Slender-billed Curlew, which has been seen by some of our interviewees, and the Ivory-billed Woodpecker, may soon be deemed to be extinct too. During the past few years, however, the number of recognised bird species has increased as species have been 'split' into two or more separate species on the basis of morphology or, increasingly, DNA evidence. As a result, today's birder is faced with more species of bird to see, even though many species have gone extinct.

Over the years it has become technically easier to see birds, and see lots of birds – what a shame that so many birds have grown rarer during that period. Foreign travel was unusual, if there weren't a war on, in the childhoods of our older interviewees, but is now much more likely to be part of an ordinary family's year.

Phil Hollom went to the Coto Doñana on something that really could be called an expedition, but now we can hop on a plane to Seville and be seeing a Spanish Imperial Eagle within five hours of leaving the ground at Heathrow (if we are lucky!).

There are better books, better optics, the internet, apps and more knowledge available to today's birder. Identification skills have certainly improved greatly, at least in those birders who care deeply about such things, but do today's birders know as much about birds as previous generations did? And do they care about birds as much?

Has the general cessation of egg-collecting by young boys, since it was made illegal in 1954, thinned the ranks of those who take up birding as a hobby? We think not, though it may mean that those who do take it up as a pastime are less well-informed about birds' behaviour and nesting habits.

The future world of birding

There will always be people who are keen to see birds – of that we are sure. Will the trends of the past fifty years of easier, quicker and cheaper travel, whether in the UK or abroad, be maintained in an era of climate change and a growing world population? We cannot tell.

We would expect technology to continue to improve, so that more and more technical aids are available to tomorrow's birders. Will they have binoculars that identify the birds for them? It doesn't seem impossible. Would we want that ability, or would it take the fun out of birding?

If trends continue, then it is tempting to think that the birder of a few decades' time will know more and more about where to go and what to see and how to identify it, but this will all be against a background of bird extinctions and declines. Will there be – perhaps there is already – rivalry to have seen the longest list of now formally extinct birds?

It's possible (and there are signs of it already) that the most environmentally aware birders will eschew the temptation of dashing around the planet burning carbon and instead opt for low-carbon birding. Plenty of birders are now as keen to maintain their county list as their world list, and events such as The

Big Sit (staying in one place and recording all the birds you see and hear) have a small but growing popularity. Might 'world list per tonne of carbon burned' one day replace 'world list' as a measure of success?

But the future is a difficult country to predict. We are living in yesterday's future, and few saw birding evolving as it has today even thirty years ago.

The internet has certainly changed the everyday life of the birder. How would the future look if the internet disappeared or became a commodity that had to be bought so that it was only available to the wealthy? How would future birding be then?

If avian flu really had killed large swathes of the human global population, as some feared it would (maybe it still will), would public attitudes to birds change so much that being a birder would be socially unacceptable?

Is there any possibility that technology could make the watching and studying (and collecting) of a different group of organisms more tempting than birds? Could birding look 'old hat' compared with the new pastime of fungus-spotting, which can be done easily in your own back garden? Might 'bio-blitzing' (where as many species as possible, of all taxonomic groups, are identified at a single site in a defined period of time) replace birding?

Will birders rise up and protest against the loss of wildlife from their lives and from their planet?

We will see. But, for now, we are going out birding.

Selected Bibliography

Anon (1969) *AA Reader's Digest Book of British Birds*. AA/Reader's Digest.

Avery, M.I. (2012) *Fighting for Birds*. Pelagic Publishing.

Avery, M.I. (2014) *A Message from Martha*. Bloomsbury.

Balmer, D.E., Gillings, S., Caffrey, B.J., Swann, R.L., Downie, I.S. and Fuller, R.J. (eds.) (2013) *Bird Atlas 2007–11: The Breeding and Wintering Birds of Britain and Ireland*. BTO Books.

Benson, S.V. (1937) *The Observer's Book of Birds*. Frederick Warne.

Birkhead, T.R. (2008) *The Wisdom of Birds*. Bloomsbury.

Brunn, B. (1972) *Hamlyn Guide: Birds of Britain and Europe*. Littlehampton Book Services.

Campbell, B. and Watson, D. (1977) *The Oxford Book of Birds*. Oxford University Press.

Carson, R. (1962) *Silent Spring*. Houghton Mifflin.

Chittenden, H. (2007) *Roberts Bird Guide*. John Voelcker Book Fund.

Clegg, J. (1967) *The Observer's Book of Pond Life*. Frederick Warne.

Cocker, M. (1989) *Richard Meinertzhagen: Soldier, Scientist and Spy*. Secker and Warburg.

Cocker, M. (1992) *British Travel Writing in the Twentieth Century*. Secker and Warburg.

Cocker, M. (1998) *Rivers of Blood, Rivers of Gold: Europe's Conflict with Tribal Peoples*. Jonathan Cape.

Cocker, M. (2001) *Birders: Tales of a Tribe*. Jonathan Cape.

Cocker, M. (2007) *Crow Country*. Jonathan Cape.

Cocker, M. (2014) *Claxton – Field Notes from a Small Planet*. Jonathan Cape.

Cocker, M. and Inskipp, C. (1988) *A Himalayan Ornithologist: The Life and Work of Brian Houghton Hodgson*. Oxford University Press.

Cocker, M. and Tipling, D. (2014) *Birds and People*. Random House.

Coward, T. (1919) *The Birds of the British Isles and Their Eggs*. Frederick Warne.

Cramp, S. *et al.* (eds.) (1977–1996) *Handbook of the Birds of Europe, the Middle East, and North Africa: The Birds of the Western Palearctic* (multiple volumes). Oxford University Press.

Crossley, R. (2011) *The Crossley ID Guide: Eastern Birds*. Princeton University Press.

Darling, F.F. (1946) *Natural History in the Highlands and Islands*. Collins.

Davies, A. and Miller, R. (2010) *The Biggest Twitch*. Christopher Helm.

Del Hoyo, J. *et al.* (eds.) (1992) *The Handbook of the Birds of the World* (17 volumes). Lynx.

Evans, G. (1972) *The Observer's Book of Birds' Eggs*. Frederick Warne.

Evans, L.G.R. (1994) *Status of Rare Birds in Britain and Ireland, 1800–1990*. LGRE Publications.

Evans, L.G.R. (1996) *The Ultimate Site Guide to Scarcer British Birds*. LGRE Publications.

Fisher, J. (1966) *The Shell Bird Book*. Michael Joseph and Ebury Press.

Fitter, R. and Richardson, R.A. (1952) *Collins Pocket Guide to British Birds*. Collins.

Forrester, R. and Andrews, I. (eds.) (2007) *The Birds of Scotland*. Scottish Ornithologists' Club.

Gantlett, S.J.M. (1985) *Where to Watch Birds in Norfolk*. S.J.M.Gantlett.

Gatke, H. (1895) *Heligoland as an Ornithological Observatory*. David Douglas.

Gooders, J. (1967) *Where to Watch Birds*. Andre Deutsch.

Gooders, J. (1970) *Where to Watch Birds in Britain and Europe*. Andre Deutsch.

Hagermeijer, W. and Blair, M.J. (1997) *EBCC Atlas of European Breeding Birds*. Poyser.

Harrison, P. (1983) *Seabirds*. Croom Helm/Houghton Mifflin.

Hayman, P. (1988) *Bird Watcher's Pocket Guide*. Mitchell Beazley.

Heinzel, H., Fitter, R. and Parslow, R. (1972) *The Birds of Britain and Europe, with North Africa and the Middle East*. Collins.

Holland, J. (1959) *Bird Spotting*. Littlehampton Book Services.

Hollom, P.A.D. (1952) *The Popular Handbook of British Birds*. H.F. & G. Witherby.

Hollom, P.A.D., Porter, R.F., Christensen, S. and Willis, I. (1988) *Birds of the Middle East and North Africa*. T. & A.D. Poyser.

Hosking, E. (1970) *An Eye for a Bird*. Hutchinson.

Jonsson, L. (2009) *Birds: Paintings from a Near Horizon*. Princeton University Press.

Kauffman, K. (2006) *Kingbird Highway*. Houghton Mifflin.

Kirkman, F.B. and Jourdain, F.C.R. (1935) *British Birds*. Nelson.

Lack, D. (1954) *The Natural Regulation of Animal Numbers*. Oxford University Press.

Lack, D. (1966) *Population Studies of Birds*. Oxford University Press.

Lack, D. (1968) *Ecological Adaptations for Breeding in Birds*. Methuen.

Leigh-Pemberton, J. (1966) *Garden Birds*. Ladybird.

Marchant, J., Prater, A. and Hayman, H. (1986) *Shorebirds – An Identification Guide to the Waders of the World*. Croom Helm.

Miller, R. (2009) *Birds, Boots and Butties 1: Anglesey*. Llygad Gwalch Cyf.

Millington, R. (1981) *A Twitcher's Diary*. Blandford.

Moss, S. (2003) *Blokes and Birds*. New Holland.

Moss, S. (2004) *A Bird in the Bush: A Social History of Birdwatching*. Aurum Press.

Moss, S. (2010) *The Bumper Book of Nature*. Square Peg.

Moss, S. (2011) *Wild Hares and Hummingbirds: The Natural History of an English Village*. Square Peg.

Mountfort, G. (1958) *Portrait of a Wilderness*. Hutchinson.

Newton, I. (1972) *Finches*. Collins.

Newton, I. (1979) *Population Ecology of Raptors*. T. & A.D. Poyser.

Newton, I. (1986) *The Sparrowhawk*. Poyser.

Newton, I. (1998) *Population Limitation in Birds*. Collins.

Newton, I. (2003) *The Speciation and Biogeography of Birds*. Collins.

Newton, I. (2013) *Bird Populations*. Collins.

Nicholson, E.M. (1927) *How Birds Live*. Williams and Norgate.

Obmascik, M. (2004) *The Big Year*. Doubleday.

Oddie, W. and Moss, S. (1997) *Birding with Bill Oddie*. BBC Books.

Ogilvie, M.A. (1975) *Ducks of Britain and Europe*. Poyser.

Peterson, R.T., Mountfort, G. and Hollom, P.A.D. (1954) *A Field Guide to the Birds of Britain and Europe*. Collins.

Prater, A.J., Marchant, J.H. and Vuorinen, J. (1977) *Guide to the Identification & Ageing of Holarctic Waders*. British Trust for Ornithology.

Rogerson, S. (1947) *Our Bird Book*. Collins.

Seebohm, H. (1901) *The Birds of Siberia: A Record of a Naturalist's Visits to the Valleys of the Petchora and Yenesei*. J. Murray.

Sharpe, R.B. (1898) *Sketch-book of British Birds*. Society for Promoting Christian Knowledge.

Sharrock, J.T.R. (1976) *The Atlas of Breeding Birds in Britain and Ireland*. T. & A. D. Poyser.

Shrubb, M. (2007) *The Lapwing*. Poyser.

Simms, E. (1971) *Woodland Birds*. Collins.

Simms, E. (1975) *Birds of Town and Suburb*. Collins.

Snow, D.W. (1958) *A Study of Blackbirds*. George Allen & Unwin.

Spirhanzul, J. and Burke, E. (1974) *Spotting Birds: A Pocket Guide to Bird Watching*. Hamlyn.

Tyler, S.J. and Ormerod, S.J. (1994) *The Dippers*. T. & A.D. Poyser.

Vesey-Fitzgerald, B. (1953) *British Birds and their Nests*. Ladybird.

Wallace, D.I.M. (1979) *Discover Birds*. Andre Deutsch.

Wallace, D.I.M. (2004) *Beguiled By Birds*. Christopher Helm.

Watson, D. and Campbell, B. (1965) *The Oxford Book of Birds*. Oxford University Press.

Watson, E.L.G. (1959) *What to Look for in Winter*. Ladybird.

Watson, E.L.G. (1960) *What to Look for in Autumn*. Ladybird.

Watson, E.L.G. (1960) *What to Look for in Summer*. Ladybird.

Watson, E.L.G. (1961) *What to Look for in Spring*. Ladybird.

Wernham, C., Toms, M., Marchant, J., Clark, J., Siriwardena, G. and Baillie, S. (eds.) (2007) *The Migration Atlas: Movements of the Birds of Britain and Ireland*. Poyser.

White, G. (1789) *The Natural History of Selborne*. Cassell & Company.

Winter, S. (2010) *Tales of a Tabloid Twitcher*. New Holland.

Winter, S. (2011) *Birdman Abroad*. New Holland.

Witherby, H.F., Jourdain, F.C.R., Ticehurst, N.F. and Tucker, B.W. (1938–41) *The Handbook of British Birds*. H.F. & G. Witherby.

Wood, J. Duncan (2003) *Horace Alexander: Birds and Binoculars*. Sessions.

Wynne-Edwards, V.C. (1962) *Animal Dispersion in Relation to Social Behavior*. Oliver & Boyd.

Index

Kirkman, F.B. 132, 208, 226
Kite, Red 87, 95, 116, 180
Kittiwake 92, 136
Kiwi 192–3
Krebs, John 87, 162, 163–4
KwaZulu-Natal, South Africa 192

Lack, David 130–1, 133–5
Lady Amherst's Pheasant 28
Ladybird books 117, 185, 208, 209
Lakenheath Fen, Norfolk 106
Langholm Moor, Scotland 139–40
Langslow, Derek 89
Langston, Rowena 89
Lapwing 50, 93, 139, 154
Lark, Bar-tailed Desert 173
Lark, Black 118
Lark, Short-toed 38, 74
Lawton, Sir John 94
lead poisoning 99–103
Lean, Geoffrey 22
Leighton Park School, Reading 208,
    209–10, 211, 213–14, 215–16
Lemsford Springs, Hertfordshire 28
Lerwick, Shetland 59, 203, 204–5
Lesvos, Greece 204
Liljefors, Bruno 222
Lincolnshire 120, 121, 123, 133, 146, 147
Lincolnshire Naturalists' Union 147
Lincolnshire Wildlife Trust 120, 147, 155
Linnet 31, 98, 131, 161, 173
linocuts 211, 218–20
Lister-Kaye, John 13
Livett, Arthur 30
Llanfairfechan, North Wales 74, 184
Loch Garten, Scotland 72, 198
Lockley, Ronald 209
The Lodge, Sandy, Bedfordshire 27, 164,
    220
Loire Valley, France 87, 89
London 18, 19, 61, 68, 109–10, 113, 174,
    176, 183

London Bird Report 61
London Natural History Society 61, 113,
    116
Loon, Common 199
Loretto School, Musselburgh 59–60
Lovegrove, Roger 152
Lucas, Derek 158
Luton, Bedfordshire 27–8
Lynch, John 19–20
Lyndhurst, Hampshire 89–90, 148

Macaw, Hyacinth 191
McCarthy, Michael 22
Machin, Graham 77
Mckinney, Frank 214
Magpie 158, 211, 218
Magpie, Azure-winged 143
Makepeace, Peter 71
Maker, Pete 28, 29
Malaysia 107, 189
Maldives 118
Mandela, Nelson 83
Manners, J.G. 4
Marchant, John 88
marine protected areas 169
Marquiss, Mick 135
Martin, House 14, 17, 178
Martin, Sand 27
Maynard-Smith, John 134
Meiklejohn, Maury 60
Meinertzhagen, Richard 54, 117
Meinypil'gyno, Russia 104–5
Mellow, Brian 28, 29
Merlin 1, 75
Michael, George 36
Middle East 16–17
Milford, Pete 44
Miller, Ruth 180–94, 225, 226
Millington, Hazel 43
Millington, Richard 31–2, 38, 40, 42–4,
    47
Milne, Brian 61